ger.com/series/6230

More information about this series at http://www.spri

Ron Welters

Towards a Sustainable Philosophy of Endurance Sport

Cycling for Life

Springer

Ron Welters
Institute for Science in Society, Faculty of Science
Radboud University
Nijmegen, The Netherlands

ISSN 1387-6678 ISSN 2215-0323 (electronic)
Library of Ethics and Applied Philosophy
ISBN 978-3-030-05293-5 ISBN 978-3-030-05294-2 (eBook)
https://doi.org/10.1007/978-3-030-05294-2

Library of Congress Control Number: 2018964574

This Springer imprint is published by the registered company Springer Nature Switzerland AG.
The registered company address is: Gewerbestrasse 11, 6330 Cham, Switzerland

Acknowledgements

While—thanks to science and technology—the impact of humankind on our planet has become pervasive, omnipresent and irreversible, our sense of agency and our ethical and political capacity to contain and govern these developments seem uncannily limited. I argue that an appropriate concept of endurance sport, in particular cycling, can help to break the deadlock between widely shared sense of urgency and discomfort and concrete action.

This book is the result of a long process, which started at the 2009 conference of the *International Association for the Philosophy of Sport* in Seattle, where I presented a hesitant first draft of my ideas on endurance, cycling and sustainability. Ever since I am enthralled by the philosophy of sport or, in the present case perhaps better, the philosophy of sustainable endurance sport. The present study is an accumulation of personal sport philosophical debris pushed by the glacier of truly great thinking. The biggest problem was to put the, during long-distance runs or cycling sessions often (seemingly) perfectly rolling, proto-philosophical ball-bearing into proper words. During the iterative process of writing, the ball-bearing always turns out to have its burrs.

Earlier versions of Chaps. 2, 3 and 5 have been published elsewhere (Welters 2014 (2 & 3), 2016 (5)), while a pre-study of Chap. 6 will be published in due time. For this study, they have been reworked, adapted and completed. Chapter 7 is the provisional result of a long-term project started by Yunus Tuncel and David Kilpatrick, whose characters can best be described with the oxymoron 'friendly agonists'. The two final paragraphs of Chap. 8 are a translation and an adaptation of parts of my ponderings on my formative years in a small village in the south of the Netherlands (Welters 2011, 2015).

After 'Seattle', further thoughts on Loland's ecosophical work, Sloterdijk's ascetological perspective and Van den Berg's metabletic critique of automated human movement practices and endurance sport in general as of 2010 have been presented and discussed during conferences of the *International Association for the Philosophy of Sport,* the *European Association for the Philosophy of Sport*, the *British Philosophy of Sport Association* and the biannual *Czech Philosophy of Sport Conference*. Many thanks to all these wonderful and self-willed sport philosophers I met there, even those with a specific interest in ballgames and/or analytic philosophy.

My special thanks for the constructive comments of and discussions with my supervisors Hub Zwart, Laurens Landeweerd and Arno Müller. I also would like to express my gratitude to the *Institute for Science in Society* at *Radboud University Nijmegen*, which patiently hosted my meandering research for many years. Especially, the lively discussions in the research group *Visions of Nature* have contributed to this study. Furthermore, many thanks for the feedback I received from a wide range of colleagues, co-racers, or, on quite a few occasions, a combination of both, in random order: Pieter Lemmens, Luca Consoli, Martin Drenthen, Leen Dresen, Paul van den Broek, Rob Lenders, Pieter Aben, Jack Niessen, Hugh Trenchard, Claus van Aanhold, Raymond Schikhof, Christoph Lüthy, Eric Dücker, Jan Vorstenbosch, Koen Breedveld, Johan Steenbergen, Riyan van den Born, Ivo van Hilvoorde, Jan Vorstenbosch, Klaas Landsman, Marjet Derks, Sander Turnhout, Leo Rouw, Jeroen Otten, Fred van Lent, Rob Cuppen, Lev Kreft, Jernej Pisk, Joca Zurc, Jésus Ilundáin-Agurruza, Javier Lopez-Frías, Sigmund Loland, Dagmar Dahl, Albert Piacente, Cesar Torres, Iris Roggema, Bep Kuling, John William Devine, Cathy Devine, Ivo Jirásek, Filip Kobiela, Irena Martinková, Jim Parry, Pam Sailors, Heather Reid, Carwyn Jones, Alun Hardman, Jon Pike, Emanuele Isidori, Alfred Archer, Paul Davis, Peter Hopsicker, Doug Hochstetler, Kenneth Aggerholm, Roeland Smits, José Luis Pérez Triviño, John Aalbers, Ellen Mulder, Imara Felkers, Alberto Carrio Sampedro, Feike Beunk, Angela Schneider, Hidde Bekhuis, Steen Nepper Larsen, James Clifford Greening, Aldo Houterman, Guus Heijnen, Jan de Leeuw, Sandra Meeuwsen, Nathanja van den Heuvel, Justus Beth, Silje van Duivenbode, Sarah Teetzel, Paul Sars, Han Rouwenhorst, Tamba Nlandu, Henk Willems, Lisa Edwards, Kevin Krein, Vera Jansen, Jerzy Kosiewicz, Patrick Hofman, Tom Vogels, Teije van Prooije, Anneloes Blom, Annechien Langeloo, Andraz Bole, Irenej Brumec, Igor Kralj, Klemen Mikic, Martina Tomsic, Noortje ter Berg, Patrick Welsch, Roald Verhoeff, Yunus Tuncel, Gunnar Breivik, David Kilpatrick, Joos Philippens, Mike McNamee, Marc Van den Bossche, Eman Hurych, Tim Elcombe, Anders Sookermany, Jeff Fry, Leslie Howe and Andrew Edgar.

I also would like to thank the students at Radboud University Nijmegen who took part in my philosophy of sport classes and the students I supervised together with Laurens Landeweerd and Lev Kreft in the *Radboud Honours Academy* Thinktank on sport. By way of lively discussions, during and after class, they often provided me with surprising and refreshing insights into e-sports and Formula 1 racing, as well as into free climbing, BMX (bicycle motocross) and mountain running. I am also highly indebted to many in-depth conversations with fanatic/frenetic fellow endurance athletes of a certain age, the so-called age groupers—rarely winning races but always striving for a personal best and qualitative improvement.

Special thanks, finally, to my supporting and loving parents Els and Hans (†), my similarly cycling and thinking brother Sven, his wife Fabienne, his son Samuel (even though he calls the little pushbike I gave him a motorcycle), my extensive family in law, and, of course, Nicolet, my ultimate cycling, running and life-mate.

Reaching a summit: Nicolet and the author in the French Alps. (Own collection)

References

Welters, R. 2011. Van Eyserbosweg naar Eyserbosweg: Fietsfilosofische plaatservaringen. In *Filosofische beschouwingen over plaats: Verbondenheid met natuur en landschap*, ed. R. Van den Born, M. Drenthen, P. Lemmens, and T. Van Slobbe, 51–70. Zeist: KNNV.

———. 2014. Sport and the environment: Ecosophical and metanoetical intersections. In *The Bloomsbury companion to the philosophy of sport*, ed. C. Torres, 163–179. London: Bloomsbury.

———. 2015. Toevende kannibaal. In *De adelaar van Benidorm: Over bijnamen in de sport*, ed. A. De Winde, O. Ibsen, S. Merlevede, and P. Verstraten, 301–305. Antwerpen: Houtekiet.

———. 2016. On ascetic practices and hermeneutical cycles. *Sport, Ethics and Philosophy* 10 (4): 430–444.

Contents

Chapter 1
Prologue: The Good Life, Asceticism and Sustainable Cycling

Bicycles let people move with greater speed without taking up significant amounts of scarce space, energy, or time.... They can get the benefits of technological breakthroughs without putting undue claims on the schedules, energy, or space of others.

(Ivan Illich 1974, p. 63)

Abstract What is the good life? Or: how are we to live? Since ancient times the answer to this question usually is that we must work on ourselves and improve ourselves by way of training. This practical and practiced philosophical investigation will focus on one particular dimension of this striving for human perfection by means of 'asceticism' (a derivative from the ancient Greek *askēsis*, meaning exercise or training): endurance sports, such as long distance running, cycling and triathlon. These are all sports that flourish by dedicated training rather than sheer motor talent, which makes them not only accessible but also increasingly popular among the crowd.

Especially the phenomenon of cycling has brought endurance sport within reach of the masses. Almost everyone can ride and afford a bicycle, a high tech artefact, which according to Ivan Illich "outstrips the efficiency of not only all machines but all other animals as well" (Energy & equity. Harper & Collins, New York, 1974, p. 60). This energetic economy makes the bicycle a straightforward tool for a more sustainable lifestyle. But the stakes of a life that is to be fully lived in endurance are higher. How can endurance sport at large and cycling in particular contribute not only to self-knowledge, but also to self-improvement and to sustainability?

Because of its competitiveness and agonistic characteristic—at first sight the very opposite of peaceful sustainable coexistence—sport usually has a negative connotation in environmental philosophy or 'ecosophy' (a contraction of ecology and philosophy), a term coined by Arne Naess in the seventies and applied to sport by Sigmund Loland in the nineties. Inspired by Loland's attempt to sketch an ecosophy of sport, and strengthened by Peter Sloterdijk's analysis of man as an upwardly oriented training animal, set forth in *You must Change your Life: On Anthropotechnics*

R. Welters, *Towards a Sustainable Philosophy of Endurance Sport*,
Library of Ethics and Applied Philosophy 37,
https://doi.org/10.1007/978-3-030-05294-2_1

(2009/2013), as well as insights from historical phenomenology (or 'metabletics'), hermeneutics and pragmatism, I will argue for a vertically challenged life in what William James has called 'the strenuous mood': serious and hard pushing, instead of pedalling around just a little. This results in an upwardly oriented ecosophical life, leading to qualitative growth, human flourishing, durability and a change for the better. Agonistic sport and environmental sensitivity: the twain shall not only meet but merge into a strenuous consequential truth.

Keywords Asceticism · Cycling · Ecosophy · Human flourishing

1.1 Towards an Ascetological-Ecosophical Understanding of Endurance

At the very heart of this interpretative-pragmatical research lies the attempt to reconcile aggressive and opposing sport and the peaceful striving for coexistent ecological sound practices. How can the idea of winning, record-setting and progress—paramount in elite sport, but to some extent usually also present in mass sport—theoretically and practically be reconciled with the ideal of peaceful coexistence of all creatures great and small, which is crucial to environmental philosophy?

In the academic philosophy of sport so far relatively little attention has been paid to sustainability and sport. The most sophisticated attempts to over-think the increasingly problematic relation between sport and environment have been undertaken by the Norwegian sport philosopher Sigmund Loland, in what I would call his 'ecosophical efforts' (1996, 2001, 2006). Ecosophy is a term coined by his fellow-countryman Arne Naess, cum founder of the deep ecology movement in the early 1970s. As indicated, it is a contraction of philosophy and ecology, and denotes a "personal total view on the relations between ourselves and nature, which guides our decisions" (Loland 1996, p. 71).

In his essay *Outline of an Ecosophy of Sport* (1996) Loland argues that in the ecological movement competition usually is considered to be the "very counter-principle to ecologically sound practice in which cooperation and symbiosis are considered key values" (p. 70). The Norwegian ecological activist Nils Faarlund, for instance, reasons that the competition-motivated lifestyle presupposes 'losers', and thus is un-ecological or non-sustainable *per se*.

> Self-realisation for the elite presupposes that the others are denied self-realisation. Competition as a value is thus excluding and elitist. Outdoor life in the sense of exuberant living in nature[1] presupposes on the other hand the self-realisation of others to achieve one's own (i.e., a presentation of self which does not separate the individual from nature).... Enjoyment in the quality of one's personal life conduct is an autotelic experiencing of value, of inner motivation (Cited in Loland 1996, p. 70).

[1] This refers to '*Friluftsliv*', the typically Norwegian predilection for unpolished outdoor-life.

Notwithstanding the problematic relation between excluding and 'agonistic' competition on the razor's edge and including and peaceful 'ecological naturalism', based on the work of Naess, Loland develops a set of hypotheses and norms that gives philosophical evidence for the idea that sport can be ecologically justified. Key terms in this interactive system of fundamental normative questions and adequate answers are 'Self-realization!'[2] on the one hand, and the idea of 'biospheric egalitarianism', or 'the democracy of all life forms', on the other. These are strivings which often collide, since human self-realisation often seems to be at odds with the self-realization of other species. Where humans appear, other species usually get the worst of it. When we really are willing to reconsider our prominent position on planet earth, we should better take a few steps back.

> For example, we might, under certain circumstances, reach the conclusion that human beings ought to recommend their own withdrawal as the dominant life form on earth to promote other life forms to live and blossom; this withdrawal may contribute more to the Self-realization for all. Such norms, critics could argue, cannot serve as a common basis for society at large (Loland 1996, p. 76).

Living a sustainable sporting life therefore equals oscillating between the Scylla of self-interest and the Charybdis of flourishing life at large. As an escape route for navigating between the two extremes Loland proposes Naess's slogan of 'simplicity in means, richness in ends' as an 'ecosophical' rule of thumb. This implies choosing the simpler, less-sophisticated option—say a standard racing bicycle rather than a high end time trail machine[3]—to satisfy ones deeply felt sporting needs. At the same time, nevertheless, technology can contribute to the ecosophical joy of sport and thus serve the process of Self-realization. For instance, the introduction of 'carving skis', which make turning more efficient and enable skiers to maintain their speed, unlike older techniques that require more refined motor skills, has made skiing easier, therefore accessible to a larger audience, and thus augmenting the net sum of ecosophical joy through augmented self-development. In short, technology in sport has a Janus-face, Loland reasons:

> [W]e can establish meaningful relations to technology, and technology can expand our experience of the unity and diversity of life. If increase in joy is able to outweigh on a long term-basis the ecological costs of production and application, sport technology can be ecologically justified (p. 84).

In his ecosophical work, Loland usually focuses on Olympic sports, especially 'record sports',[4] such as the 100 metre dash in athletics. For good reasons he argues that these sports represent the (in the end untenable, un-sustainable and technology-driven) 'record sport dilemma': "*The continuous quest for new*

[2] With a capital S and an exclamation-mark, indeed.

[3] In 2017 bicycle manufacturer Cervélo launched the P5X as the 'ultimate tri bike', which costs approximately 15,000 euro's.

[4] "A sport record is a performance, measured in exact mathematical-physical entities (meters, seconds, or kilograms) within a standardized spatio-temporal framework defined by sport rules, that is better than all previous performances measured in identical ways" (Loland 2001, p. 128).

records is built on the impossible quest for unlimited growth in limited systems" (2001, p. 130).

Other than Loland's focus on fast fibres and spectacular elite athletics,[5] in this study I will draw special attention to endurance sports, such as swimming, cycling, long distance running, triathlon (a mixture of these three disciplines) and duathlon (a.k.a. run-bike-run). I will argue that these are from a spectator's view perhaps less attractive, but they have special benefits to offer when it comes to sustainability issues. They mirror the logic of non-linear qualitative growth.

In Loland's terminology endurance sports are 'quasi-record sports', since "they are measured and compared by exact timing, but there are no standardized arenas" (2001, p. 128).[6] Ironman distance triathlons,[7] for instance, can be performed on 'fast' courses, such as in Roth, Germany. There it took the men's winner in 2016 7:35:39 (a new 'world record')[8] to cross the finish line. But there are also triathlons which take place under extremely rough conditions. Consider, for instance, the *Isklar Norseman Xtreme Triathlon* in Norway. The winner's time in 2016 was 10:22:37 (Norwegian Lars Petter Stormo). This means that the Norseman is relatively slow, but extremely difficult. This specimen of *über*-endurance starts with swimming in an ice-cold fjord. After being dragged out of the water by special auxiliary troops, who help competitors to strip off their wetsuits,[9] they have to cycle through a stunning but barren mountainous landscape. The ordeal is rounded off

---

[5] The 100 metre dash is considered the highlight of the modern Olympic Games: explosive, athletic, highly concentrated, and often performed by athletes with great commercial value and charisma. At the 2016 Olympics in Rio de Janeiro sprinter Jamaican Usain Bolt, for instance, received more exposure and media coverage than Kenian Eliud Kipchoge, the winner of the men's marathon. While at top level both disciplines require a lot of natural talent, in endurance sport training, perseverance, dedication, steadiness and *phronēsis* (practical wisdom), are also essential characteristics, which, I argue, have special benefits to offer for a full-blown ecosophical take on sport.

[6] The third category of sports Loland mentions is constituted by 'games', such as soccer, tennis and handball. There seems to be another, fourth category, however: so-called juried-sports, such as gymnastics and platform diving. One might furthermore question if combat sports (e.g. boxing, mixed martial arts, free fighting) are a mixture of games or contests and juried sports or a separate category In the more methodologically en theoretically oriented Chapters (especially Chap. 5) I will address the problematic issue of categorising human activities people are inclined to call 'sport' in more detail, among others by reconsidering Bernard Suits's tricky triad of games, play and sport (1988).

[7] An Ironman triathlon consists of 3,8 kms of swimming, followed by 180 kms cycling and rounded with running a marathon (42.195 kms).

[8] It is debatable to use the term 'record' here, since there are no standardized arenas in triathlon. Triathlon is a quasi-record sport in Loland's grammar: "The Boston Marathon is rather different from the one in Oslo. The conditions and trails of cross-country ski races vary from race to race. We talk of records here but in an accurate way. Events with exact performance measurements but without strictly standardized frameworks shall be referred to as quasi-record sports" (Loland 2001, p. 128).

[9] Due to severe cooling most competitors can't do this by themselves. To my knowledge, in all other long distance triathlons help from outside is forbidden. Unlike in professional cycling, competitors in a triathlon are supposed to be self-supporting, except for the so-called 'energy labs', where the athletes can stock up on food and fluids.

with a marathon that ends with a frightening climb to the infamous Gaustatoppen, a.k.a. 'Zombie Hill' (elev. 1,883 m.), according to the organisers "a massively steep slope creeping up to the beginning of the rugged trail run towards the peak". During the unpaved part of the climb, which consists of 7 km steep upwards over boulders, competitors are obliged to be accompanied by two helpers, who can offer assistance in case of injury or total exhaustion.

> The athletes took to the mountains posthaste, climbing out of the valley and into the clouds to cross Hardangervidda in some of the most intense and challenging weather ever seen on Norseman. Thick gusty fog, biting rain and 5°C summer weather could deter the most hardened survivor, but the athletes fought through it—unfortunately, not without casualties (Meyr 2016).

Norseman: Finishing at Zombie Hill. (Source: Norseman Xtreme Triathlon)

At this point the take on sport as a playful and voluntary attempt to overcome unnecessary, self-raised obstacles (Suits 1978/2005) does not seem to apply anymore. For a different outlook and a broader perspective on the wavering ways of *homo faber*[10] I will therefore propose, criticise and refine the work of the 'post-Heideggerian' and 'post-Nietzschean'[11] German philosopher Peter Sloterdijk (1947–). Throughout this book, in each chapter from a different angle, I will apply his findings on the ways of

[10] A term coined by Karl Marx, meaning 'man the maker'. In this study *homo faber* will reveal him- or herself as a *homo asceticus*, an inevitably practicing man or woman. In human life it is not about the things that are made, it is about the making—so process rather than result—a point which often seems to be overlooked in sport philosophy (Cfr. Aggerholm 2016).

[11] Friedrich Nietzsche and Martin Heidegger are Sloterdijk's main philosophical role models. Overall one might call his style eclectic, exuberant and distinctive, however. A more comprehensive reading of Sloterdijk's philosophical style will be provided in Chap. 5 *Ascetic Practices, Hermeneutical Cycles and Ecosophical Endurance.*

human beings as natural born practitioners in search for a meaningful life fully lived in endurance, especially cycling.

With regard to the inconvenient ecological truth which, according to a plethora of scientific findings, we are currently facing, Sloterdijk argues[12] for the development of so-called 'homeotechnologies', a neologism[13] referring to the ancient Greek *homeos*, meaning 'alike or 'similar'. Other than allotechnology (*allo* means 'other' or 'alien' in Ancient Greek), the development and implementation of more ecosystem-friendly technology may ideally lead to a 'homeotechnical turn', which, according to the Dutch environmental philosopher Sanne van der Hout, provides "specific opportunities for a more peaceful co-existence of humans and nature" (2014, p. 423).

Van der Hout argues, nonetheless, that Sloterdijk tends to overestimate the "rescue potential of homeotechnology" (2014, p. 437). Elaborating on his ideas unfolded in *You must Change your Life: On Anthropotechnics* (2009/2013), during a lecture at the United Nations Climate Conference in Copenhagen in 2009 entitled *How Big is 'Big'?,* Sloterdijk has argued that people quite probably:

> will remain convinced that it is the task of evolution through constant growth to globalize material prosperity and the expressive privileges they themselves enjoy. They will refuse to come to terms with a future that is based on reduction and restraint (cited in Van der Hout 2014, p. 428).

Sloterdijk apparently does not seem to believe in the ability of humans to give up their 'kinetic' and energy-consuming lifestyle. Or, if we measure his findings on the ecosophy-scale: humans tend to strive for quantity and growth rather than quality and restraint. Thus: homeo-technology is the best we can get.

The acknowledgement of the ubiquitous kneeling of mankind for shallow hedonism might suggest that Sloterdijk opts for a technological fix of ecological problems (*adaptation*) rather than a radical change (*mitigation*) of our hyper-consuming lifestyle. In this narrowing perspective new and clean technology is supposed to resolve the problems caused by traditional polluting industry. Van der Hout agrees that "thanks to the emergence of homeotechnology the resilience[14] of the earth can be increased" (2014, p. 433). She warns, however, about the pitfall of a totalitarian homeotechnical control over nature, since, paradoxically, following nature may even open up new horizons for genetic manipulation.

---

[12] Roughly as of his *Spheres Trilogy*, published together in 2005 as *Being and Space,* a reference to Heidegger's *Being and Time* (1927/2008).

[13] Sloterdijk shares Heidegger's habit of enriching the already richly worded German language with witty but often somewhat enigmatic linguistic findings. He does so, however, with a more light-hearted touch, I contend.

[14] In the Chaps. 6, and particularly 7, I will raise the issue of how to increase the physical resilience of humans, so that they will take less refuge with an unsustainable life-setting. A physically more active life may result in less polluting ways of travelling and probably also to a more food conscious consumption: "less meat more veggies" (Peter Sloterdijk in Giesen 2011). From a moral point of view it seems reasonable to not only shift the blame to our vulnerable planet, but to take responsibility ourselves. Of course developing homeo-technology seems fair, but on the other hand we should not stop questioning our own behaviour, restrain ourselves and ponder over the idea of an 'appropriate withdrawal', which will be taken up further on.

> What's more, since current technoscience obscures the classical distinction between 'biomachines' and 'manmade machines', this exploitation runs the risk of becoming increasingly subtle and invisible. Thus, homeotechnology may result in strengthening our sway over nature even on a molecular level (p. 425).

This dystopic warning against a brave new biotechnological world seems morally appropriate. From a historical empirical viewpoint, however, one can hardly neglect the human inclination towards progress by means of ever more sophisticated and pervasive technology. History learns that people tend to go upwards, not just onwards. On a larger philosophical scale the multiple characters of our inescapable upwardly oriented strivings—our almost congenital[15] 'vertical challenge' in Sloterdijk's ironic wording—also has been acknowledged by authors such as the Dutch science-philosopher Hub Zwart (2016):

> Several continental philosophers (from Hegel, Schopenhauer and Nietzsche up to Binswanger and Sloterdijk) have thematised this persistent drive towards increased complexity in terms of the vertical dimension of existence, a dimension which is most obviously noticeable in humans. Human beings, these authors claim, not only strive for continuation, propagation and reproduction (the horizontal dimension), but also for optimization and self-improvement (both individually and collectively), through exercise, experimentation and technology (p. 254).

The idea of exercise of all sorts, rather than technology as a means for qualitative human progress and flourishing,[16] is omnipresent in Sloterdijk's *You must Change your Life*. By concentrating on physical exercise as a specific form of 'self-technique'—according to Sloterdijk implying "all forms of self-referential practicing and working on one's own vital form" (2009/2013, p. 340)—I will argue that the 'anthropotechnical' turn promoted by him offers a well-operable escape route for real 'ecosophical' change, beyond the calculative limits of homeo-technology.

In the course of this argument for a sustaining concept of sport, the idea of humans as practicing and necessarily "upward-tending" animals (p. 13) will reveal itself as a practical-philosophical blueprint for a life lived in appropriate moderation. This ideally results in a rather 'simple' life that, nevertheless, can optimally flourish, in the sense that growth towards well-spirited quality is to be preferred over hedonistic quantity. If well-understood and well-performed, asceticism teaches us that the good life comes with a substantial effort.

When put to the ecological litmus-test, the idea that we must learn to revalue our potential to work on ourselves—the central message of Sloterdijk's provocative *You must Change your Life*—almost automatically leads to the insight that we should invest in our stamina. We should swim, run or hop on a bicycle rather than in a car

[15] In *Driven by Technology. The Human Condition and the Biotechnology Revolution* (2008) Pieter Lemmens, furthering Sloterdijk's findings, contends that technology is constituent of the human condition since the very beginning. Humans are essentially technological, not accidentally.

[16] As expressed in Aristotle's concept of εὐδαιμονία (*eudaimonia)*, best translated as 'well-spiritedness'.

or a plane. I will argue that the bicycle has special benefits to offer when it comes to an ascetical ecosophical change for the better. [17]

The more physical side of the practicing life usually still has a rather negative connotation in academic philosophical spheres, however. While 'the body' or 'physicality' as such over the last two decades has been re-introduced in philosophy by authors such as Wilhelm Schmidt and Michel Onfray, the devotedly and intensely training and (competitively) sporting body still is largely ignored. A deep frontal wrinkle still prevails over a sweaty face. This seems to be the case because of the supposedly aggressive, ruthless and record-setting (thus un-ecosophical) disposition of sport.

This preference for the cerebral is however undeservedly so, since competitiveness (even on the cutting edge) historically preceded the noble art of methodical thinking. Or, as the Turkish-American sport philosopher and Nietzsche-expert Yunus Tuncel (2013) stipulates: philosophy originally was drenched in an atmosphere of disputative rivalry. He concludes that in ancient Greece the agonal[18] spirit "had already existed in poetry, mythology, arts, and *athletics* (italics mine), before it re-produced itself in thought with the rise of philosophy" (p. 255). In a similar vein Sloterdijk argues for a more physically oriented basic attitude towards the still

---

[17] It has to be noted that José Ortega y Gasset's ideas on asceticism and technology, already formulated in the nineteen-thirties, look similar to Sloterdijk's, who doesn't mention the Spaniard's work in *You must Change your Life* once. However, there is at least a difference of emphasis in the interpretation of asceticism of both authors. Whereas Sloterdijk paves the way for the beneficial 'democratising' wider implications of ascetology, in *The Revolt of the Masses* (1932/1993) Ortega seems at least hesitant about the beneficial range of true asceticism for all.

As one advances in life, one realises more and more that the majority of men—and of women—are incapable of any other effort that that strictly imposed on them as a reaction to external compulsion. And for that reason, the few individuals we have come across who are capable of a spontaneous and joyous effort stand out isolated, monumentalised, so to speak, in our experience. These are the select men, the nobles, the only ones who are active and not merely reactive, for whom life is a perpetual striving, an incessant course of training. Training = *askesis*. These are the ascetics. This apparent digression should not cause surprise. In order to define the actual mass-man, who is as much 'mass' as ever, but who wishes to supplant the 'excellent', it has been necessary to contrast him with the two pure forms which are mingled in him: the normal mass and the genuine noble or man of effort (pp. 65–66).

In other words: mass asceticism always bears the traits of reactive and external 'horizontality'. Jesús Ilundáin-Agurruza elucidates that (like Sloterdijk's) Ortega's ratio-vitalistic asceticism comes with a substantial vertical effort. "This ethos embraces challenges that demand suffering and discipline" (2014, p. 289). In the end Ortega y Gasset's asceticism remains in noncommittal 'ludic' spheres, however, Ilundáin-Agurruza contends: "It also values activities non-instrumentally, for what they are, focusing on the process not the result" (p. 289). For Ortega sport is "the most exalted, serious, and important part of life, while labour ranks second as its derivative . . . life, properly speaking, resides in the first [i.e., sportive activity] alone; the rest is relatively mechanic and a mere functioning" (Ortega y Gasset 1941/2002, p. 18). Whilst also Sloterdijk is anything but a result fetishist, I still argue that his general vertical ascetology is more susceptible for a massive behavioural change for the better, since it implicitly also aims for concrete results: a more sustainable world. It is bottom-up mass (endurance) sport that will change our life for the better, rather than ('ludic', playful) top-down elitarianism.

[18] *Agon* means, among other things, 'fight' 'or 'struggle' in Ancient Greek. Further on I will elaborate the concept of *agon*, notably in Chap. 7.

cerebrally fixated, reflective 'love for wisdom'. He also does so by referring to the physical basis of ancient Greek culture.

> The word 'philosophy' undoubtedly contains a hidden allusion to the two most important athletic virtues, which enjoyed almost universal popularity at the time of Plato's intervention. It refers firstly to the aristocratic attitude of 'philotimy', the love of *timè*, that glorious prestige promised to victors in contests, and secondly to 'philopony', the love of *pónos*, namely effort, burden and strain (p. 194).

You must Change your Life is to be considered a variegated attempt to explore the wider meaning of asceticism, specifically in its physical manifestation, and to use its lingering transformative potential for a change for the better. Sloterdijk argues that a well-performed and properly understood ascetic (and therefore active) life may lead to a personal *metanoia*—a turning point, a radical reform, a mitigation of a consumptive, hedonistic and superficial lifestyle. On a collective societal level asceticism even may lead to a 'renaissance', which in Sloterdijk's exuberant philosophical grammar is not just bound to fourteenth century Italy, but marks a broader variety of massive shifts in preferential ascetic practices.

Sloterdijk deducts from a vast empirical stock that each era has its own specific training regimen. These shared programs attract at first sight scattered, but, all things considered, analogous practices. As of the introduction of the modern *Olympics* in 1896 the 'athletic ideal'—so prominent in classic Greek culture, with its already mentioned preference for physical *agon* and ditto *askēsis*—has taken the lead again. According to Sloterdijk this is a "transformation best described as a re-somatisation or de-spiritualisation of asceticisms" (2009/2013, p. 27).

> In sport, the spirit of competitive intensification found a almost universally comprehensible, and hence globally imitated, form of expression. It not only completed the 'rebirth of antiquity', but also provided the most concrete illustration of the performative spirit of modernity, which is inconceivable without the de-spiritualization of asceticisms. De-spiritualized asceticism is known as 'training', and corresponds to a form of reality that demands fitness as such, fitness *sans phrase*, of individuals (p. 335).

At this decisive point the key note of this research comes to the forefront once again. How to reconcile aggressive and adversarial sport and the peaceful striving for coexistent ecological sound practices? Or, put in Sloterdijk's baroque wording: how to integrate personal *metanoetic* strivings—the quest for individual fitness—into a collective and unifying 'renaissancistic', re-birthing[19] framework?

At present the rapid ecological degradation of Planet Earth should be our main concern, Sloterdijk argues. One way to face environmental challenges is to train

[19] Initially rebirthing just referred to a type of curative breathing technique invented by Leonard Orr, who proposed that correct breathing can cure disease and relieve pain. I contend that this ode to the benefits of proper breathing techniques perfectly fits in Sloterdijk's suggestion for a physically oriented ascetological framework. Things definitely went wrong, however, when Orr devised rebirthing therapy in the 1970s after he supposedly re-lived his own birth while in the bath. His claim that breathing techniques could be used to purge traumatic childhood memories that had been repressed has been thoroughly criticized and unveiled as unscientific and only attractive for an irrational clientele by Margaret Singer and Janja Lalich in *Crazy Therapies: What are they? Do they Work?* (1996). In 2006, a panel of over one hundred experts participated in a survey of psychological treatments considered rebirthing therapy to be discredited.

ourselves to become better citizens, contributing to a just and sustainable society. In order to do so we *must*[20] change our lives and strive for a new horizon of universal co-operative training practices. This results in a comprehensive 'general ascetology' which, if properly understood and well-performed, paves the way for a better common future. However, as already argued, even those philosophers who pretend to embrace the more physical side of human life, so far have largely neglected sportive physical training techniques as a driving force in human life.[21]

> One can only do justice to the Modern Age as a whole if one relates it to a mental, moral and technological change that has never been adequately portrayed: the existence of the moderns shows aspects of a global fitness exercise in which what I have termed the 'ethical distinction', the intense call to elevate life—heard by very few in premodern times—has been transformed into a universally addressed and multifariously answered metanoetic imperative (pp. 335–336).

Sloterdijk's provocative findings regarding our inborn craving for heavy training practices, and their inherent transformative potentiality for a change for the better, opens up a more positive evaluation of our upwardly oriented sportive strivings than Faarlund suggests in his characterization of (competitive) sport as non-sustainable *per se*. Or to put this informed claim in bicyclical terms: "Anyone can fight on flat stretches, but those who remain capable of fighting a duel on the worst of mountains already deserve to be called Hector or Achilles."[22]

1.2 Philosophical Accounts Beyond Play

Peter Sloterdijk's imaginative use of cycling imagery is not only a conspicuous metaphor; it is also an actually existing alternative means for a change for the better. Therefore the 'push-bike'[23] as a high-tech but still relatively 'clean' preferential tool

---

[20] 'Must' (*musst*) is not very common in written German, in which there is a strong preference for the more indirect 'should' (*sollst*).

[21] This critique is also shared by the Belgian philosopher Marc Van den Bossche (2010), who concludes that the rise in contemporary attention for 'philosophy and the art of living' usually does not include the sweaty ways of what Meinberg (1991) has coined *homo sportivus*.

[22] Thus spoke Sloterdijk in an interview concerning the Tour de France (Gorris and Kurbjuweit 2008).

[23] 'Pushbike' is a British-American expression I came across during a coast to coast cycling trip in the US in 2001. This expression distinguishes the bicycle from a motor-cycle, on which one just has to sit, and not to push the pedals. It is the lifting of the knee and the pressure on the calves and quads that makes the difference between supposedly feeling free and truly being alive. While both the motor-cyclist and the push-cyclist have developed a routine of photographing themselves next to sign that indicates the height of a climb, their physical experience of the way to the top is totally different: stepping on the gas versus intensely pedaling with sore calves and quads. Other than Robert M. Pirsig suggests in his novel *Zen and the Art of Motor-cycle Maintenance* (1974), I argue that riding a push-bike is the best way to actually understand the United States in the footsteps of Lewis & Clark. I will take up this argument in the final Chapter, entitled *Epilogue: Turning in the Widening Gyre.*

for carving out an appropriate niche for ecosophical flourishing will be gleaming behind the scenes throughout this study. This will be punctuated with first and second hand endurance sport experiences, again, especially those of the cycling kind. It takes an effort—no *philotimy* without *philopony*—but once sufficiently trained, every philosophical cyclist will experience William James's magnanimous sensation of a second wind (1907).

A closer investigation and casuistic application of Sigmund Loland's, Sloterdijk's and other philosophical thoughts on sustainable sport, often still *in statu nascendi*, will result in a blueprint for an 'ecosophical ascetology' (Welters 2014). This training program for substantive ecological improvement and for a profound mitigation of our over-consuming life-style, paves the way for a durable conception of endurance sports—from elite athletes, dedicated age-groupers and 'finishers'[24] up to occasional joggers and weekend-warriors (Hochstetler and Hopsicker 2012). The polishing of a broad-angled lens with sufficient depth of focus on sport in a deteriorating natural setting will be the main objective of Chap. 2, entitled *Sport and the Environment: Considering Sustainable Thoughts.* In Chap. 3, *Answering Three Ecosophical Questions: Asceticism*, I will put my ecosophical considerations to the test.

In Chaps. 4, 5 and 6 I will argue for a more holistic, synthetic and processual[25] view on sustainable endurance sport. This integrative and deepening tripartite effort is a necessary exercise, since contemporary academic sport philosophy is often locked into the binary view of narrow and broad internalism. Narrow internalists, or formalists, argue that sports are solely constituted by their rules. This restrictive approach can be characterized as the 'autotelic' or self-referential stance. In this view sport is (just) organised and rule-bound play that is supposed to remain in its well-defined own domain. Sport is (or ought to be) nothing but sport. Because of the formal and discursive approach of its subject matter, narrow internalism is said to have strong connections with analytic philosophy.[26]

Broad internalists, or interpretivists, on the other hand, reason that sport is more than just a 'lusory' (playful or gratuitous) end in itself. This widening perspective has strong bonds with (but, as I will argue, is not entirely to be reduced to) philosophical hermeneutics, and advances the idea that sport is not just sport, but also can be interpreted as a full-fledged means towards other ends. For instance, power, prestige and prize-money on the more cynical side, and a deeper knowledge of life,

---

[24] In long distance endurance sport events many 'competitors' do not aim for a high ranking, they just want to finish the race within the time limit.

[25] The Danish sport philosopher Kenneth Aggerholm (2016) has undertaken a well-designated effort to link Sloterdijk's ascetological findings with a plea for a less result-oriented and more process-oriented virtuous approach in the philosophy of sport. Other than his generalist and still quite 'ludic' approach, I will argue for a more specific endurance sport oriented interpretation of ascetic 'aretism', or virtue, however. I will elaborate on this in Chapter 8.*Epilogue: Turning in the Widening Gyre.*

[26] See for the problematic character of the supposed watershed between 'analytic' and 'continental' philosophy Fusche Moe (2014), whose findings in this will be addressed in more detail in Chap. 5 *Ascetic Practices, Hermeneutical Cycles and Ecosophical Endurance*.

human flourishing, and even experiencing Immanuel Kant's notion of the sublime on the idealistic side of the spectre.

I will elaborate on the claims of broad internalism by respectively putting historical phenomenology (or 'metabletics'), hermeneutics, and pragmatism to the ascetological-ecosophical litmus-test. In Chap. 4, entitled *Metabletics of Spinal Sport: When Poion meets Poson,* the historical phenomenology—or 'metabletics'—of the controversial[27] Dutch psychiatrist and phenomenologist Jan Hendrik van den Berg (1914–2012) will be operationalised to grasp the subject matter at stake: The viability of (endurance) sport, specifically cycling, as a tool for carving out meaning beyond the often-cited portable definition of Bernard Suits of game-playing[28] as a "voluntary attempt to overcome unnecessary obstacles" (1978/2005, p. 55).

Metabletics is the 'doctrine of change'.[29] It mirrors the conviction that humans really are different, over time and in varied settings. The methodological baseline is the idea that there are striking similarities between at a first glance unrelated events occurring in a certain period. *De Reflex* (1973) is a metabletic critique of contemporary egalitarianism and thoughtless and uncritical herd behaviour. Van den Berg wants to switch on consciousness again and make humans real, alive persons again, who are more than just a bundle of unreflective reflexes. He is horrified by the post-religious celebration of sheer physical acts that, literally, run over our spine, without the faintest interference by the cortex. The virulent author wants to take us back to, literally, pre-reflexive times. Van den Berg particularly disapproves of those who practice sport in a group. Regarding cycling in a peloton he sighs: "The reflex discerns herb nor tree, knows no nature; the reflex has no land, no homeland" (117, my translation).

Counter to Van den Berg's intuitive rejection of spinal sporting practices, I will change the perspective in the direction of a more positive attitude towards seemingly pointless training and competitive practices. My metabletic analysis will focus on the miracle year of 1974, a highlight in the history of sport, arguing that 1974 is for the greater part a year of *poion*, of quality, of fully being in the world. Whilst, by comparison, 2010 turns out to be instead a year of *poson*, of quantitative, Narcissistic self-orientation.

---

[27] Van den Berg may be considered a radical conservative with sometimes questionable, and anything but egalitarian thoughts. Still, for the same reason one might argue that his provocative thoughts open up challenging new societal horizons, for instance when it comes to our apparent inclination to turn towards technological fixes, rather than bridling ourselves and returning to a life that is less consumptive—a striving which also often shows up in 'leftist' eco-philosophy, with its incompatible . In this sense Van den Berg's critique echoes Arne Naess's plea for human modesty and the anti-hedonistic undercurrent in his 'non-anthropocentrism'. I will elaborate on this in Chap. 4 *Metabletics of Spinal Sport: When Poion meets Poson.*

[28] The philosophy of sport suffers from a semantic confusion when it comes to game-playing and sport Suits himself speaks of a 'tricky triad' consisting of play, games and sport. 1988. This will be discussed in more detail in Chap. 5 *Ascetic Practices, Hermeneutical Cycles and Ecosophical Endurance.*

[29] *Metaballein* means 'to change' in ancient Greek.

Nevertheless, furthering Van den Berg's metabletic spinal intuition towards an ecosophical-ascetological understanding is still worthwhile, I argue. A proper and deeper understanding of the repetitive nature of endurance sport, might even pave the way for a radical turn in life on the individual and the collective level. Once engrained through dedicated training practices the reflex will even turn out to be the very necessary condition for truly experiencing the land, perhaps even a homeland. Reflexes ideally result in reflections.

Chapter 5, *Ascetic Practices, Hermeneutical Cycles and Ecosophical Endurance,* embellishes broad internalist approaches which recently have been undertaken recently in the philosophy of sport,[30] and focuses on endurance sport in particular. Self-improvement through repetitive practice (*askēsis*) has always been a key theme of human existence. Our 'ascetic planet' is inhabited by individuals who are constantly and relentlessly training themselves for the better. This may seem self-focused, but it may also have a broader scope: we train ourselves to become better humans, thereby contributing to a just and sustainable society. Paradoxically, however, this will only work when we become fully aware of our exercises as forms of life that engage the practicing person. For accomplished bicyclists means and ends, and narrow and broad outlooks on the training and sporting subject, will result in what Sloterdijk refers to as the sensation of being submerged in a "spinning cycling pantheism" (Gorris and Kurbjuweit 2008).

An applied and specified hermeneutics of endurance cycling can enrich our understanding of this sports activity as a form of asceticism. As such I will elaborate a view on cycling as an upwardly oriented 'spiral' that can contribute not only to self-knowledge and self-improvement on the individual level (*metanoia*), but also to an 'ecosophical renaissance' on the collective societal level. Real ascetics are prone to go upwards, not just onwards. Those who learn to agonize in order to optimize their ascetic potentialities will in the end become better human beings, step by step (Welters 2016).

From the application of Sloterdijk's dialogical hermeneutics we learn that virtuous, upwardly oriented spiral cycling is a preferential form of asceticism that can be performed in the great wide and preferably hilly open spaces at your own pace. By cultivating the good habit of cycling as a relatively sustainable 'anthropotechnical' means we not only learn to fight the struggle with and against ourselves, but also to transform our often uncompromising attitude towards our natural environment into a coexistent ecosophical agony with the world. Somehow we are all natural ascetics born into a world filled with hindrances. Thus, we may argue that in a meaningful ascetic life cycling is an involuntary attempt to overcome necessary obstacles.

For metanoetic inspiration Sloterdijk in *You must Change your Life* also briefly refers to William James's (1842–1910) thoughts on religious conversions (pp. 304–306). The idea of betterment through well-understood and well-performed training programmes has a stronger connection with pragmatism in general and James's Spartan plea for the 'strenuous mood' and 'precipitousness', however. Also James's

[30] A special issue of *Sport, Ethics and Philosophy* has been solely dedicated to sport and hermeneutics (Lopez Frias and Edgar 2016).

description of the blessing of experiencing a beneficial 'second wind', unravelled in *The Energies of Man* (1907), that one may experience after intense training practices of all sorts, implicitly resonates in Sloterdijk's plea for asceticism.

> The rhythms of regeneration hold the secret of the overexertion that leads to higher performance levels. This phenomenon has been intuitively comprehensible since time immemorial, and had already been exploited for intensive training sessions in antiquity; on the other hand, the ancients were also familiar with the phenomena of overtraining that appear if regeneration rhythms are disregarded. With increases in mental and fine motor performance, supercompensation is augmented by a form of superadaptation. ... Recent research in the fields of learning theory, neuro-motorics, neuro-rhetoric and neuro-aesthetics consolidate and vary didactic intuitions that originate in early asceticisms and artistries (2009/2013, pp. 320–321).

Sloterdijk does not shrink from hyperbole and extremes to score his ascetological point. He also refers to Franz Kafka's hunger artist for instance, who continued to starve himself to death, even when the audience had forgotten he existed.

> Like the hunger artist, the athletes have a message for the psychologically poorest and the physically weakest that is worth sharing in: the best way to escape from exhaustion is to double the load (p. 417).

In Chap. 6, *Continental Pragmatism: Enduring Life in the Strenuous Mood* (2012), the pragmatist outlook on endurance sport, proposed by Douglas Hochstetler and Peter M. Hopsicker and inspired by William James, will be taken up and put into perspective. I agree with the authors that to fully flourish, we have to become true runners and real cyclists rather than dabbling joggers or occasional weekend warriors indeed. Counter to their claim for determining the golden mean between sportive achievement and social environment (2016), I will argue for a more individualistic and perhaps even somewhat anti-social view on the Aristotelian ideal of human flourishing, or *eudaimonia*, however Decidedly developing ones very own personal 'vertical challenge' may be the incentive for the more sustainable lifestyle we need on the collective level—the highly necessary ecosophical renaissance. Only those who climb mountains—preferably on their own, or in a small group, and at their own pace—deserve to be called a Hector or Achilles, showing the way towards a more sustainable lifestyle by acting accordingly.

Picking up the pieces, and trying to further equalise the meandering and bumpy uphill path towards meaning and consequential truth, enables us to cast a look at the stunning scenery while climbing, instead of watching for potholes at every single step or pedal-stroke. Van den Berg's historical phenomenology has proven its benefits as a social critique of 'spinalism'.[31] But in the end metabletics fails to understand 'spinalized' human movement practices as the very possibility of overcoming the oblivion of being truly human. It seems as if Van den Berg overlooks the true

---

[31] I propose this generic term for a criticism of our high tech and literally inhumane society which largely overlaps with Heidegger's *Seinsvergessenheit* ('forgetfulness of being' or 'oblivion of being'), as well as to some extent Sloterdijk's already mentioned criticism of allo-technology, and the more frequently used 'critique of technical reason' (E.g. Van den Bossche 1995).

nature of spinal exercises.[32] Hermeneutics, the second theoretical impetus brought to the fore, turned out to be capable of defending verticality against allegations of being unsustainable by putting asceticism in a larger historical interpretive framework. These continental and broad internal assets so far have resulted in an ascetic plea for cycling uphill at your own pace as an involuntary anthropotechnical attempt to overcome necessary hilly obstacles.

Nevertheless, there still remains a non-operational metaphysical nucleus in both depicted 'continental' initiatives for developing a broad internal view on the desirable ways of *homo asceticus*. How to move up from subtle and contextualized deep thinking to concrete actions? How to convert the call from the stone to change our lives for the better into feasible deeds? To counter this heuristic flaw, during the philosophical-theoretical underling of my argument I have introduced a rudimentary version of pragmatism to strengthen and materialise my claims for a full and sustainable life in endurance. In the final part of the theoretical chapter *Continental Pragmatism: Enduring Life in the Strenuous Mood* (Chap. 6) I will tentatively further the (potentially beneficial) pragmatist take on endurance sport. Particularly William James's radical empiricism and teleological pragmatism—as expressed in catch-phrases as 'truth in consequences' and 'truth's cash value'—will prove its added value. His subtle and well-worded pragmatism[33] may even overcome the classic philosophical stalemate between, successively, perfect ideas and imperfect shades, *ousia* and *psyche*, universals and particulars, *res cogitans* and *res extensa*, *noumena* and *phenomena*, realism and relativism, *noesis* and *noema*, *signifiant* and *signifié*, things and words, and all other infertile and in the end inoperable dichotomic attempts to grasp being or the problematic status of the world. According to James practices cannot be understood with tools, things, ideas or words that exist outside these practices. It is the practice itself that constitutes meaning.

---

[32] This is an explanatory weakness metabletics shares with 'classical' phenomenology. The attempt to get to the core of things paradoxically leads to alienation of the matter in question. The Dutch philosopher of technology Laurens Landeweerd stipulates that the famous dictum of the founding father of phenomenology Edmund Husserl (1859–1938), *'Zu den Sachen Selbst'* (towards the things themselves) "is notorious because for those who follow him, his phenomenological ideas seem to imply a reduction of knowledge of things to knowledge of appearances" (2016, p. 1). "In stressing the textu(r)al nature of our reality, and analyzing the underlying structures of our understanding of reality as more fundamental than that reality itself, the above schools of thought tend to reduce things to contextu(r)ality, thus robbing us of the ability to experience their 'thingly' nature. As such, they echoed the death of metaphysics (as the philosophy of being) and its replacement by the discursive, something already premeditated by Ludwig Feuerbach. As early as 1848, he stated that his age held 'a preference for the sign to the thing signified, the copy to the original, fancy to reality, the appearance to the essence'" (p. 11).

[33] Not be confused with 'crude pragmatism' in sports coaching: "the view that the right approach to coaching is the one that gets results', most obviously on the scoreboard or on the clock" (Devine and Knight 2017, p. 35).

> Things and thoughts are not at all fundamentally heterogeneous, they are made of one and the same stuff,[34] *which* as such cannot be defined [ontologized] but only be experienced; and which, if one wishes, one can call the stuff of experience (James 1976, p. 271).

Integrating the benefits of phenomenology, hermeneutics, and pragmatism, I will opt for what the sport-minded Belgian philosopher Marc Van den Bossche has titled a "Heideggerian pragmatism" (1995, pp. 128–141).[35] Furthering this post-ontological and "meta-philosophical relativist" (Van den Bossche 2001, p. 44) synthesis enables to challenge the prevailing word view of humans as sporting for either just prize money, playfulness or health, by 'philoponically' working up the holistic idea of (endurance) sport as a preferred manifestation of "the art of living" (Van den Bossche 2010, p. 137).

Perceived through the resulting broad-angled lens, not only things and thoughts but also concrete acts of the human trainer will merge into what Sloterdijk in *You must Change your Life* depicts as a "multi-disciplinary and multi-virtuosic world with expanding limits of ability" (2009/2013, p. 155). In the last two chapters, then, I will revitalise the findings I gathered so far on the subject matter at stake—leading the good life in well-understood and well-performed ecosophical ascetic practices—in this potent continental pragmatic vein.

In Chap. 7, entitled *On Agon and Ecosophical Endurance: Finding your own Pace*, I will review my former tentative attempts to reconcile agonistic sport with co-existentially oriented 'ecosophy' in relation to an in-depth analysis of endurance sport practices. I will do so, once again, with a special focus on the big why of the strenuous life on two un-motorized wheels. Riding a bike does not require much natural talent or motor ability, nor is cycling highly damaging for our (due to a sedentary and seated life-style) increasingly fragile joints.[36] This makes the bicycle fairly easy accessible for the crowd. Beyond this rather shallow salutary argument for enhancing bicycle use, however, I will focus on the issue of the still increasing amount of endurance 'athletes'[37] that run, cycle or swim "far beyond a healthy jaunt" (Gorichanaz 2016, p. 367), often even without the slightest chance of winning a race. This indeed comes close to the dictum, wrongly ascribed to the found-

---

[34] Prospero, the right duke of Milan, already argued in a similar tone of voice on the impossibility of separating things from thoughts or nominal words from universal ideas: "We are such stuff as dreams are made on; and our little life is rounded with a sleep" (Shakespeare 1983, p. 19).

[35] Van den Bossche argues that pragmatism is not exclusively an Anglo-Saxon case. Heidegger's philosophical toolbox also accommodates pragmatic chisels. Cfr. also Okrent (1988), Drabinski (1993) and Pollock (1995) for other attempts to integrate the phenomenological-hermeneutical tradition with pragmatism. I will elaborate on this line of reasoning further on, in particular in Chap. 7 *On Agon and Ecosophical Endurance: Finding your Own Pace.*

[36] Unless one crashes or rides knee-devastating gears, of course.

[37] 'Athlete' usually refers to a natural born talent for sport, in particular the athletic disciplines, that can be enhanced by but are not totally reducible to training practices. I argue that in long distance endurance sport genetic predisposition is far less important than the willingness to train over and over again. It is the true ascetic attitude that counts.

ing father of the Modern Olympics, Pierre de Coubertin: taking part in the games is more important than to win.[38]

An analysis of narratives of dedicated practitioners of all sorts—from elite athletes to age group competitors and just finishers of long distance endurance ordeals—will reveal that those who learn to really agonize themselves in order to optimize their ascetic potentialities will become better humans in the end, with or without winning medals. Tough uphill cycling—competitive or not, alone, in a small group or in a peloton—is a preferential form of asceticism that can be performed in the great, wide (and preferably hilly) open spaces. Pursued at your own pace and understood as a process of gradual and 'spiral' self-improvement, rather than as beating opponents, endurance sport overcomes the stalemate between serious sport and sustainable strivings. I argue that only the strenuous, vertical life that pays full tribute to *pónos*, and enjoys the pleasure of experiencing a second wind, or the blessing of super-compensation after long periods of nearly exhaustion, is worth to be lived. The call for asceticism turns out to be a metanoetic imperative for bi-pedalism.

> Bicycles let people move with greater speed without taking up significant amounts of scarce space, energy, or time. They can speed fewer hours on each mile and still travel more miles in a year. They can get the benefits of technological breakthroughs without putting undue claims on the schedules, energy, or space of others. They become masters of their own movements without blocking those of their fellows. Their new tool creates only those demands which they can also satisfy (Illich 1974, p. 63).

Therefore: hop on your bike; conquer the world at an appropriate pace in a sustainable manner. Improve your life, find meaning and inspire others to follow the same steep, winding and sometimes bumpy road.

In Chap. 8, entitled *Epilogue: Turning in the Widening Gyre*, I will reflect on the ecosophical-ascetological line of reasoning I set out on and pick up the pieces of a life lived in endurance, especially of the cycling kind. Starting off with challenging gravity and the first glorious unaided pedal strokes in the backyard. Followed up by widening circles: villages, boroughs, provinces, countries, continents. And then, in reverse order again, in search of the origin (Welters 2011). On city bikes, racing bikes, time trail machines, mountain bikes and sturdy trekking bikes. On paved and unpaved surfaces. Alone or in a small group. Commuting, travelling, racing, or just riding. Finding one's own pace and understanding that the good ascetic life comes with un-motorized automated moving, specifically cycling. "Non-racers. The emptiness of those lives shocks me" (Krabbé 1978/2016, p. 1).

---

[38] Cfr. Loland (1995): "Coubertin's perhaps best known aphorism is actually a quote from a bishop of Central Pennsylvania, Ethelbert Talbot, who, during a pulpit sermon during the 1908 London Games claimed that '...the important thing in these Olympiads is less to win than to take part in them'"(p. 64). On the same page Loland concludes that Coubertin's real concern was the sporting record, the 'external axiom' of Olympianism. This winning mood is perfectly exemplified in the Olympic motto *Citius, altius, fortius* (faster, higher, stronger). The Games are about winning.

Cyclists and cars: Never the twain shall meet. (Own collection)

1.3 User's Manual

Sisyphus, the king of Ephyra (now known as Corinth), was punished by Zeus for his self-aggrandizing craftiness and deceitfulness by being forced to roll an immense boulder up a hill, only for it to roll down when it nears the top, repeating this action for eternity.

This book represents the final episodes of a lived-through ascetological process with a 'quasi-Sisyphean' character. Writing about an endurance sport is an endurance activity in itself. It is a long and winding ascent with steep intersections and detours. It is a trail-run with unforeseen hindrances. I realise that this represents some challenges for readers as well. Sometimes, the same mountain slope comes into view, but from a different perspective. Sometimes, the terrain may seem too strenuous, but nonetheless it must be covered.

This is a long-distance thesis, and reading it may require some capacity for endurance. Each chapter of this book is a training session with its own story, starting off with a warm-up and followed by a ride or race on which sometimes a suddenly appearing side path is taken, to see if there is anything to experience. Some parts of the course are covered several times, albeit from different approach routes and under varying weather conditions. At the same time, these preparatory runs and rides are also stages with an ultimate goal constantly in view: to understand that the good and sustainable life comes only after a substantial vertical effort.

References

Aggerholm, K. 2016. On practising in sport: Towards an ascetological understanding of sport. *Journal of the Philosophy of Sport* 43 (3): 350–365.

Devine, J.W., and C.J. Knight. 2017. Beyond 'crude pragmatism' in sports coaching; Insights from C.S. Pierce, William James, and John Dewey, A commentary. *International Journal of Sport Science and Coaching* 12 (1): 35–37. https://doi.org/10.1177/1747954116684405.

Drabinski, J. 1993. Radical empiricism and phenomenology: Philosophy and the pure stuff of experience. *Journal of Special Philosophy, New Series* 7 (3): 226–242.

Fusche Moe, V. 2014. The philosophy of sport and continental philosophy. In *The Bloomsbury companion to the philosophy of sport*, ed. C. Torres, 52–65. London: Bloomsbury.

Giesen, P. 2011, April. Peter Sloterdijk, Duitse filosoof: Naar de sportschool om de planeet te redden. *De Volkskrant* 30 (2011): 7–8.

Gorichanaz, T. 2016. Beautiful and sublime: The aesthetics of running in a commodified world. *Journal of the Philosophy of Sport* 43 (3): 365–380.

Gorris, L., and Kurbjuweit, D. 2008. Philosopher Peter Sloterdijk on the Tour de France: "The Riders are Just Regular Employees." *Der Spiegel Online International*. http://www.spiegel.de/international/europe/philosopher-peter-sloterdijk-on-the-tour-de-france-the-riders-are-just-regular-employees-a-565111.html. Accessed 9 September 2017.

Heidegger, M. 1927/2008. *Being and time*. New York: Harper & Row.

Hochstetler, D., and P. Hopsicker. 2012. The heights of humanity: Endurance sport and the strenuous mood. *Journal of the Philosophy of Sport* 39 (1): 117–136.

———. 2016. Normative concerns for endurance athletes. *Journal of the Philosophy of Sport* 43 (3): 335–350.

Illich, I. 1974. *Energy & equity*. New York: Harper & Collins.

Ilundáin-Agurruza, J. 2014. José Ortega y Gasset: Exuberant steed. *Sport, Ethics and Philosophy* 8 (3): 285–314. https://doi.org/10.1080/17511321.2014.981361.

James, W. 1907. *The energies of man*. The American Magazine (reprint).

———. 1976. *The works of William James: Essays in radical empirism*. Cambridge: Harvard University Press.

Krabbé, T. 1978/2016. *The rider*. London: Bloomsbury.

Landeweerd, L. 2016. *Das verborgene Leben der Dinge/The hidden life of things*. Unpublished lecture/Artistic Object.

Lemmens, P. 2008. *Gedreven door techniek. De menselijke conditie en de biotechnologische revolutie (Driven by technology. The human condition and the biotechnology revolution)*. Oisterwijk: Box Press.

Loland, S. 1995. Coubertin's ideology of Olympism from the perspective of the history of ideas. *Olympika: International Journal of Olympic Studies* 4: 49–77.

———. 1996. Outline of an ecosophy of sport. *Journal of the Philosophy of Sport* 23: 70–90.

———. 2001. Record Sports: An ecological critique and a reconstruction. *Journal of the Philosophy of Sport* 28: 127–139.

———. 2006. Olympic sport and the ideal of sustainable development. *Journal of the Philosophy of Sport* 33: 144–156.

Lopez Frias, J., and A. Edgar. 2016. Sport and hermeneutics. *Sport, Ethics and Philosophy* 10 (4): 343–349.

Meinberg, E. 1991. *Die Moral im Sport, Bausteine einer neuen Sportethik*. Aachen: Meyer & Meyer.

Meyr, K. 2016. *Isklar Norseman extreme triathlon: Results 2016*. https://nxtri.com/wet-wet-cold/. Accessed 5 February 2018.

Okrent, M. 1988. *Heidegger's pragmatism. Understanding, being, and the critique of metaphysics*. Ithaka: Cornell University Press.

Ortega y Gasset, J. 1932/1993. *The revolt of the masses*. New York/London: W.W. Norton.

———. 1941/2002. The sportive origin of the state. In *Toward a philosophy of history*, ed. J. Ortega y Gasset. Urbana/Chicago: University of Illinois Press.
Pollock, A. 1995. *Pragmatism and hermeneutics: Rorty, James and Gadamer on truth*. Ottawa: University of Ottawa.
Shakespeare, W. 1983. *The complete works of William Shakespeare*. London: Henry Pordes.
Singer, M., and J. Lalich. 1996. *Crazy therapies. What are they? Do they work?* San Francisco: Jossey-Bass.
Sloterdijk, P. 2005. *Sphären: Eine Trilogie*. Frankfurt am Main: Suhrkamp.
Sloterdijk 2009/2013. *You must change your life: On anthropotechnics*. Cambridge: Malden.
Suits, B. 1978/2005. *The grasshopper: Games, life and utopia*. Ontario: Broadview Press.
———. 1988. Tricky triad: Games, play, and sport. *Journal of the Philosophy of Sport* 15: 1–9.
Tuncel, Y. 2013. *Agon in Nietzsche*. Milwaukee: Marquette University Press.
Van den Berg, J.-H. 1973. *De reflex: Metabletische tegelijk maatschappijkritische studie*. Nijkerk: Calllenbach.
Van den Bossche, M. 1995. *Kritiek van de technische rede. Een onderzoek naar de invloed van techniek op ons denken*. Utrecht: Jan van Arkel.
———. 2001. *Ironie en solidariteit. Een kennismaking met het werk van Richard Rorty*. Rotterdam: Lemniscaat.
———. 2010. *Sport als levenskunst*. Rotterdam: Lemniscaat.
Van der Hout, S. 2014. The homeotechnical turn: Sloterdijk's response to the ecological crisis. *Environmental Values* 23: 423–442.
Welters, R. 2011. Van Eyserbosweg naar Eyserbosweg: Fietsfilosofische plaatservaringen. In *Filosofische beschouwingen over plaats: Verbondenheid met natuur en landschap*, ed. R. Van den Born, M. Drenthen, P. Lemmens, and T. Van Slobbe, 51–70. Zeist: KNNV.
———. 2014. Sport and the environment: Ecosophical and metanoetical intersections. In *The Bloomsbury companion to the philosophy of sport*, ed. C. Torres, 163–179. London: Bloomsbury.
———. 2016. On ascetic practices and hermeneutical cycles. *Sport, Ethics and Philosophy* 10 (4): 430–444.
Zwart, H. 2016. What comes next? The operator theory as an operationalisation of the Teilhardian view on cosmogenesis. In *Evolution and transitions in complexity*, ed. G.A.J.M. Jagers op Akkerhuis, 253–267. Cham: Springer.

Chapter 2
Sport and the Environment: Considering Sustainable Thoughts

If increase in joy is able to outweigh on a long-term basis the ecological costs of production and application, sport technology can be ecosophically justified

(Sigmund Loland 1996, p. 84).

Abstract In recent years there has been much attention for environmental matters in actual sport practices, ranging from greening the Olympics to reducing the ecological footprint of mass running events. In itself these adaptive developments are praiseworthy. At the same time they raise philosophical discomfort, because they do not fully address the issue of how to mitigate the effects of our over-consuming and polluting life-style in a more profound way.

This Chapter provides a rendition of a still tender and tentative philosophical and ethical debate by iteratively attempting to bridge the gap between 'shallow' adaptive green sport practices and 'deep' eco-philosophical thinking. It does so with a special focus on Sigmund Loland's work on the 'ecosophy' (a portmanteau of ecology and philosophy) of sport. Notwithstanding the problematic relation between excluding and 'agonistic' competition on the razor's edge and including and peaceful 'ecological naturalism', based on the work of Arne Naess, Loland develops a set of hypotheses and norms that gives philosophical evidence for the idea that sport can be ecologically justified. Key terms in this interactive system of fundamental normative questions and answers are 'Self-realization!' through engaging in sporting activities on the one hand, and the idea of 'biospheric egalitarianism' and 'the democracy of all life forms', on the other. Or: how to find the right mean between the pleasure of sporting in nature and sustainability?

After analysing and criticising Loland's sensitive *Outline of an Ecosophy of Sport* (1996), I will critically access two consecutive 'sport-ecosophical' articles by Loland: *Record Sports: An Ecological Critique and a Reconstruction* (2001) *and Olympic Sport and the Ideal of Sustainable Development* (2006). Whereas Loland concentrates on developing a robust sport-ecosophical mindset and leaving the idea of sport records, since these represent the logic of unlimited growth in limited systems, I rather suggest perhaps 'shallow' but nevertheless concrete acts that actually result in a change for the ecosophical better.

R. Welters, *Towards a Sustainable Philosophy of Endurance Sport*, Library of Ethics and Applied Philosophy 37,
https://doi.org/10.1007/978-3-030-05294-2_2

Keywords Environmental issues · Self-realization · Nature · Sustainable sport

2.1 Sustainable Sport

At the1992 *United Nations Conference on Environment and Development* (UNCED) in Rio de Janeiro, environmental issues were officially placed on the world-wide political agenda. This striving for a solid basis for a sustainable world resulted in 1999 in the *Earth Summit Agenda 21*, a blueprint for a balanced and integrated approach to environmental and developmental issues into the next millennium. At the Centennial Olympic Congress (1994), the International Olympic Committee (IOC) took up the gauntlet and began to develop its own 'green' ideology. From then on next to 'excellence' and 'friendship', also 'respect' (apparently for the environment in its entirety) officially became a third pillar of Olympianism. This resulted in programmatically putting forward an environmentally conscious conception of sport and declaring the responsibility "to promote a positive[1] legacy from the Olympic Games to the host cities and host countries" (IOC 2012, p. 5). Although in the report the IOC is aware of the quite ambitious character of its sustainability strivings, the organization doesn't shun being self-confident about its ability to bring about a change for the better.

> There are those who believe that this is an impossible dream—as many doubted that the sub-1(sic![2]) minute mile or sub-10 second 100 metres would ever be achieved. However, the IOC understands that if it is to fulfil its aim to 'Create a way of life based on the joy of effort, the educational value of good example, social responsibility and respect for universal fundamental ethical principles' there is no other option (p. 94).

The idea of unbridled optimistic progress as a mean for a better world is paramount in Olympianism. The IOC even suggests to transform its ideal of athletic excellence—as expressed in the official Olympic motto of *citius, altius, fortius*: ever running faster, jumping higher and getting stronger—into a sustainability panacea. "It is to be hoped that future achievements in the drive to foster widespread sustainable development will inspire a great many people across the globe. That way we truly incorporate the ethos of Olympism—faster, higher and stronger for a sustainable future" (p. 94).

This optimistic view on sport as a progressive and potent enhancer for societal change has been paramount since the beginning of the modern Olympic movement. The idea of Olympianism as a *religio athletae*, "a secular, vitalistic 'humanism of the muscles'" (Loland 1995, p. 66), that is supposed to bring about a change for the better, already has been poetically expressed by founding father Pierre de Coubertin at the very year of the first modern Olympics in Athens in 1896.

> Healthy democracy, wise and peaceful
> internationalism, will penetrate the new stadium and
> preserve within the cult of honour and
> disinterestedness which will enable athletics
> to help in the tasks of moral education and social
> peace as well as of muscular development
> (cited in Loland 1995, p. 49).

---

[1] Which also implies 'sustainable', I reason from the context.

[2] This should be 4 minutes, referring to Roger Bannister's first sub 4 minute-mile (3:59:4) in 1954.

As a contemporary topical follow-up for De Coubertin's optimistic moral activism, as of 1994 the IOC strives for Olympic Games that are as energy neutral as possible. As a consequence of this formalised striving we could have had 'green' Winter Olympics in Vancouver in 2010. Were it not that for an unexpected lack of snow-fall in ski-resort Whistler highly energy consuming and loads of carbon emitting snow-cannons had to be flown in from all over the world to let the (in case quasi-hibernal) Games go on.[3]

Environmental responsibility meanwhile even seems to have become a common denominator in sport at large. As a consequence of this blossoming consciousness we have had a fairly green *Grand Départ* of the Tour de France in Rotterdam, also in 2010. We also have the promise that the FIFA World Cup 2022 in Qatar will take place in an eco-friendly soccer utopia. There is green golf (less herbicides), green tennis (more durable tennis balls), and green ice-hockey (low energy consumption buildings). And of course outdoor sports such as mountain-biking (don't leave litter, stay on the sign-posted tracks) and trail-running (bring your own cup to refuel at the energy stations) are often labelled as green.

Still, beyond the usual window dressing, one may question: what's in it for life at large? The clash of the praise of the wild and concrete behaviour to satisfy the human need for the pristine is epitomized by the hipster in lumberjack, dropped in the sheer endless Canadian pine-woods by a chartered helicopter. To concretize and personalize the strained relation between thoughts and acts and head and heart further, taking a look at Richard Branson seems appropriate. This venture capitalist combines running a record company not only with aircraft discounting,[4] sponsoring Formula 1 racing and promoting space travelling, but also with trans-ocean kite-surfing and calling himself a true environmentalist, unselfishly caring for the well-being of planet earth. Meanwhile even Formula 1 race car driver Lewis Hamilton has called for taking more care of the environment: "Little by little we kill our planet. We humans are not a very good species" (Bloemhof 2017, my translation). This is rather ironical, if not cynical, since one can hardly think of a sport more polluting than car racing. Who is this 'we'? What about personal responsibility, exemplary conduct and concrete action?

Branson's environmental Janus face and Hamilton's uncommitted sigh seem to confirm Martin Heidegger's (1889–1976) famous critique of *'Das Man'* (the They) in *Being and Time* (1927/2008): "The 'who' is not this one, not that one, not oneself [man selbst], not some people [einige], and not the sum of them all. The 'who' is the neuter, the They [das Man]" (p. 164). Or: "Everyone is the other, and no-one is himself" (p. 165). Who is meant by 'the They'? Everyone? The arithmetical mean of all opinions, including those on the extremes?[5] Or is 'the They' more of a *moi commun*,

[3] To quote then IOC president's Avery Brundage (USA) famous words after the murder of 11 Israeli athletes by Palestinian terrorists during the Munich summer Olympics in 1972.

[4] "Air travel has the highest specific impact on short-term warming, while on long-term warming car travel has an equal or higher impact per passenger-kilometer" (Borken-Kleefeld et al. 2010, p. 5700).

[5] In the case of Heidegger it seems as if his ideas are rather singled-out thoughts formulated against the very idea of a 'regression to the mean'. This in the social sciences widespread statistical technique has turned into the very opposite of the original intention of the founder of the statistical analysis Francis Galton, who coined his calculative analysis regression to the *mediocre* (1886). Galton was in search of *Hereditary Genius* (1892), the exceptional, not in the dull mediocre "sum of them all".

a *pluralis majestatis,* a generalised I which at a closer look represents the particular meaning of a specific person or author who refuses to take personal responsibility?

Warily hovering between the dull average mean and the brilliant special case is a major asset of practical philosophy. Other than fundamental ontology and meta-ethics, a true philosophy of the true praxis, and hence practical ethics, thrive by well-contextualized N = 1 case studies, since these have a maximum declarative potency. This is cynically captured in Stalin's famous dictum that one dead soldier is a tragedy; while a million dead soldiers are just statistics.[6] Of course, there always remains an obligation to hop over from anecdotal vividness to societal robustness. However, although necessary in practical philosophy, in the end the broader empirical turn in itself never is sufficient. I find that a practical philosopher may start with an anecdote, a singled out observation that initially raised his or her attention (often a discomforting and alienating one), subsequently has to tap the wider empirical angle, but, finally, also has to depict an overarching perspective on the issue at stake—an inviting prospect resulting in a workable high morale.

Following this initially quasi-anecdotal but finally meta-empirical and practical-ethical trajectory is the overall philosophical line of reasoning of this study. Arduously pedalling for meaning, this is the thread of this poly-pragmatic philosophical study.

2.2 Sport and Nature Under Siege

In *Sportfilosofie* (2000), a comprehensive overview of the continental and analytic philosophy of sport, Jan Tamboer and Johan Steenbergen provide an overview of the scarce attempts to think through the challenges of sustainable sport that have been undertaken so far. The authors, both with a background in sport and exercise sciences and philosophy, argue that in the ethics of sport there traditionally is a strong focus on the proper behaviour of the individual sports(wo)man (p. 183). They reason however that due to recent developments the often-complex interrelation between sport and the environment should be at the forefront of any attempt to formulate an inclusive contemporary ethics of sport.

Tamboer and Steenbergen distinguish four actual practices where the detrimental effects of sports on our natural habitat are manifest. First: the vast use of natural reserves for sport (from sports fields to stadium car parks). Second: there is direct (car racing) and indirect (travelling to matches and sports tourism) pollution. Third: the increasing pressure on and even destruction of ecologically vulnerable biotopes (e.g. the detrimental effects of skiing on flora and fauna and even on local communities). Fourth: practicing sport often causes sound pollution (yelling masses in stadia, sponsor caravans in the *Tour de France*, music support during large running or triathlon events, etc).

[6] This conviction is also uttered by Mr. Van Arkady in Paul Theroux's novel *Half Moon Street* (1984). Van Arkady argues that there are only 5,000 people in the world that really matter. The rest is literally meaningless, and thus non-existent. "I believe the death of one man can change the course of history, when it is the right man and when we are fully conscious of it. But a million don't matter, because it isn't a number in any actual sense, unless it is applied to money" (p. 7).

At the societal level Tamboer and Steenbergen confirm the growing awareness of the harmful effects of sport on the environment among sports associations mentioned before. Environmental advisors are hired and sportspeople are being environmentally educated to make sports practices sustainable, at least to some extent. On the one hand there is a movement to limit the explosive growth of 'nature sports' such as skiing, rafting and mountaineering, and thus the negative effects these outdoor sports exert on our natural habitat. From a purely practical point of view this would be salubrious for the environment as such. Less human activity simply means less environmental damage. On the other hand it should be noted that governmental measures for enhancing the health of the increasingly sedentary crowd by practicing outdoor sports are flourishing. In this line of reasoning sporting in nature is associated with health and with reconnecting with nature as such. In this positive vein sporting in nature will straightforwardly lead to respecting nature more.

With regard to this increasingly popular idealistic idea of sport as a proper tool to counter what Martin Heidegger in *Being and Time* (1927/2008) has coined "*Seinsvergessenheit*" (forgetfulness of being),[7] Edwin Jakob and Gunnar Liedtke (2004) refer to nature-sports as bearing a high conflict-potential (p. 15). They stipulate that sports like canoeing and rock-climbing not only take place in vulnerable biotopes, but due to a tendency towards more convenience and safety also often result in building facilities in those last untouched refuges—and thus cause secondary environmental damage.

When it comes to close encounters of strolling wanderers and rampaging mountain-bikers, there often seems to be a fundamental reciprocal misunderstanding between the two denominations of outdoor-lovers. To strengthen this point, I hereby present an anecdotal observation from my own experience as an endurance sport practitioner. In my hometown I observe on a regular basis the migration of the weekend masses to the woods, riding their mountain bikes for an hour or so, experiencing nature and driving back home. In the lovely medium-sized university town where I live, there also is a looming weekend-walking-squad—usually also driving (with often a preference for 4WD, I notice) and often dressed up in green (natural!) wax-coats. This grey-haired army is not amused by the, at least in their perception, reckless and aggressive armada of mountain-bikers, dressed in bright-coloured spandex (unnatural!). Some of these aggrieved lovers of nature even have developed the habit of putting small tree trunks right across those tricky parts of the official mountain-bike track. Exactly at those spots where such obstacles may cause damage. This nasty habit made me have a serious crash, a sore back and a broken front wheel once. And since I was the last one on the track just before sunset (and since I do not own a mobile phone or a driving licence—unlike my opponents dressed in dark green, I often observe) the cunning ambush arranged by the complacent walking-squad forced me to drag myself home for many kilometres.

[7] Translated to the specific case at hand: through domestication and convenience practices we have lost the ability to experience the harsh physical side of life in its unveiled bare 'natural' state. For a more detailed analysis of the suppressed agonistic side of life Cfr. Chap. 7 *On Agon and Ecosophical Endurance: Finding your own Pace.*

Heumensoord (Nijmegen, The Netherlands): A mixture of activities in nature: sheep; walking; running with dogs; relaxed mountain biking and horse-riding; speeding up a bit; the yearly competitive run-mountainbike-run. (Sources: Staatsbosbeheer & Triathlonvereniging Cerberus Groesbeek)

With just a little bit of mutual consideration we could have it all, I argue: walking the (preferably leashed) dog, running, mountain-biking, thoughtless strolling around, and even Nordic walking. The wooded area around my hometown is vast enough. In reality, there is a clash of different cultures and different perceptions of nature, however: the slow and appeased versus the fast and furious. Camouflage meets bright spandex. From the nature-friendly perspective the mountain-biking variant of the sportsperson is the alleged aggressor in the on-going battle for space. Of course there is a negative impact of mountain-biking on nature: twigs will be broken and over time tracks will wear out, but the same goes for all human activity on Planet Earth. *Homo mensura*, man is the measure of all things. In the course of this study I will elaborate the concept of a well-willing coexistent ecosophical concept of sport, which, nevertheless, still keeps up the idea of agonism.

Furthermore, instead of (just) concentrating on the supposed effects on the *loci delicti* as such (sports grounds, signposted mountain-bike tracks, buildings facilities which also often serve as after sport chill-outs) also the tendency to travel all over the world to pristine natural outdoor sporting areas has to be taken into the environmental moral sport-calculus. The all too human urge to travel to ecologically vulnerable places of interest at all costs and without delay is very harmful for the over-burdened Planet Earth.[8] Either to watch a hellish mountain stage in the *Tour de France* or to actually participate in the *Isklar Norseman Xtreme Triathlon,*[9] mentioned in the previous chapter, there is at least a sense of moral ambiguity in emitting loads of carbon-dioxide in order to passively or actively enjoy the last pristine outposts of planet earth. To find the right mean between the retrograde and participative stance is one of the major challenges when it comes to sketching a workable blueprint sustainable sport.

In *Sport, Culture and Society: An Introduction* (2012) Grant Jarvie and James Thornton take up the environmental challenge by proposing a more robust empirically oriented heuristic framework for a sustainable conception of sport by considering the specific power of sport as a transformative tool in a sociological-political vein. They distinguish radical dark green and reformist light green approaches to environmental ethics in sport. The first mindset is firmly rooted in anti-capitalism and activism, tends to be proactive and relies on the beneficial effects of deep ecology, social ecology and eco-socialism. In the dark green ideology there is intrinsic value in nature; sport is not above nature but necessarily embedded in nature. The second school of thought tends to be reactive and is more or less based on conservatism (in the literal sense), 'preservationism', stewardship, free-market liberalism and social (but free-market) reformism. This essentially implies "tradable pollution rights plus voluntary agreements plus regulation" (p. 272).

---

[8] Further on I will argue for the bicycle as an appropriate, relatively 'clean' mean to satisfy the human need for travelling.

[9] As already mentioned in Chap. 1 long distance race which ends with climbing (rather than running) the steep, unforgiving and unpaved slopes of Gaustastoppen (1883 m), a.k.a. 'Zombie-Hill'. It is compulsory for every participant to have a supporting team for the last 7 extremely steep and dangerous uphill kilometres.

The light green approach, which is predominant in current technological times, reasons that man is a well-willing master of nature. Jarvie and Thornton end with a carefully worded conclusion that says:

> Sport may not be able to halt major environmental catastrophes but its undoubted popularity in many parts of the world means that it provides a popular target for organizations such as Greenpeace, the anti-golf movement[10] and the IOC to deliver on environmental messages. Perhaps the real question for environmentalists is: can we have sport at all without nature? (p. 280).

The answer to this provocative question depends on how we interpret and frame 'nature', naturally. Do we see nature as fixed, something pristine out there, untouched and uncontrolled by humans (the hidden assumption in our un-problematic use of the lemma nature in every-day speech)?[11] Or do we frame nature as dynamic and intertwined with human activity per se? In philosophical shorthand: do we perceive nature as more or less fixed actuality or quasi infinite potentiality?

For Benedict de Spinoza (1632–1677) *natura naturata* ('nature natured') refers to nature as a passive product of an infinite causal chain. *Natura naturans* ('nature naturing'), on the other hand, denotes the self-causing activity of nature, or nature in the active sense, as Samuel Taylor Coleridge (1772–1834) has phrased the issue of how to deal with the often used but hardly ever well-defined theoretical term 'nature'. Everything changes, nothing lasts, as the ancient Greek philosopher Herakleitos (c. 535–c. 475 BC) already claimed: πάντα ῥεῖ (panta rhei). Ever newer waters flow on those who step into the same river. Or as the dictum usually is phrased: it isn't possible to step in the same river twice. In Spinoza's ethereal, pantheistic and deistically[12] connoted composition the two-sided nature of nature—fixed versus dynamic—is phrased as follows:

> [By] *Natura naturans* we must understand what is in itself and is conceived through itself, *or* such attributes of substance as express an eternal and infinite essence, that is … God, insofar as he is considered as a free cause. But by *Natura naturata* I understand whatever follows from the necessity of God's nature, *or* from God's attributes, that is, all the modes of God's attributes insofar as they are considered as things which are in God, and can neither be nor be conceived without God (Spinoza 1996, p. 26).

To put this ontological bifurcation in contemporary, less theologically driven words: God (or any other culturally determined equivalent) created nature, but not for once and for all. Within creation there is a certain openness which leaves space for

[10] Keeping golf-courses that 'unnaturally' green as they usually are takes huge amounts of herbicides.

[11] Cfr. for how an analysis of actual human visions (the so-called lay(wo)men's perspective) of nature may help to strengthen environmental philosophy e.g. Van den Born: "An empirical turn can make environmental philosophy more responsive to the views that actually exist in the wider public. Beyond a mere affirmation or rejection of the existing philosophical notions, we can discover people's own voice" (2007, p. 180).

[12] Although Spinoza never explicitly refers to deism in his writings, his philosophy shares a lot in common with the idea of God as an absent 'clock-winder', who enables life but does not steer it in a pre-determined direction. A de-spiritualized neo-Darwinian version of this image is presented in Richard Dawkins's *The Blind Watchmaker* (1987).

(positive) human intervention. God is silent *dynamis* (potentiality), nature is flourishing *energeia* (actuality).

Peder Anker's (1998) persuasive reprise of Spinoza's theological naturalism in terms of 'created' (*naturata*) versus 'creative' nature (*naturans*) is productive for our specific case that will be put to the litmus-test hereafter: sport and nature, or rather: sport ín nature.[13] "Spinoza assumes that Nature is the immanent cause of itself and of all other things, and that, as a result without limits" (p. 1). In other words: sport and nature are two sides of a coin that ideally will reinforce each other. Good natured sport wipes out its traces. Picking up this Spinozistic proto-evolutionist idea of nature as unbridled potentiality may help to overcome the unfruitful and artificial opposition of idyllic and untouched nature versus ruthless and usurping sport.

A chapter on sport ethics and environmental ethics by the German sports-pedagogue Eckhard Meinberg vents the un-convenient marriage between *homo sportivus* and environment. In 1991 he observed that until then the natural environment of sport usually was considered self-evident and unproblematic, and therefore largely anathema in mainstream sport ethics. Building on a rich German tradition of philosophical anthropology, particularly Arnold Gehlen and Helmut Plessner, Meinberg sets forth his what may be called 'Spinoza-inspired' position by arguing that:

> We always encounter non-human nature, including the natural environment, never directly, non-natural, but always in a prefabricated cultural context. … either if we want to dominate or we want to feel her and the like ….To the watchful observant of the world of sports it will be dawning that sport, not only elite sport in many of its manifestations, has become a first-rate moral provocation, and it increasingly becomes clear that lately the relation between sport and the environment is ethically dubious (1991, p. 132).

Alpine skiing is a land-consuming activity that causes deforestation,[14] and obviously motorized sports are polluting, Meinberg continues. Furthermore, notwithstanding the fact that they took up environmental issues as early as in the 1970s, 'clean' sports like kayaking and sailing cause irresponsible and irreversible negative environmental effects. Also the popular phenomenon of 'Nordic' cross-country skiing is seen as a real environmental threat, because of the fact that this activity threatens the natural habitat of animals and plants.

Meanwhile even horse riding and jogging have become environmentally suspect, Meinberg contends (p. 133). After all, these (at first glance rather bucolic) activities may harm the natural environment: breaking twigs, damaging trails, leaving litter, etc. And, once again: travelling to and from events must also be taken into account when it comes to the moral calculus of the environmental pressure of sport at large.

---

[13] Which still is possible in an environmentally positive manner, I will argue.

[14] Zwart (2017) points at the changing nature of skiing by referring to the days when Heidegger was skiing from his famous hut in Todtnauberg to the lecturing hall of the university of Freiburg: "The meaning of skiing may have changed dramatically during recent decades, however, so that nowadays skiing not only exemplifies forgetfulness of nature but also the inauthenticity of contemporary consumerist existence" (p. 2).

Given the vast circumstantial evidence, it must be concluded that the expanding pursuing of elite as well as mass sport activities has a double-edged effect: sport is increasingly becoming hostile to the environment, and, subsequently, environmentalist perspectives are becoming more hostile to sport. Meinberg therefore suggests a first draft for an ethics of sport that is environmentally conscious and tries to reduce its ever increasing ecological footstep. Such a co-existential ethics, which respects nature and sport equally, can be supported by four 'particulars', he reasons.

First, the idea of co-existence should not only be descriptive but should also be a *normative touchstone* for sporting behaviour. In this respect producing artificial snow for alpine skiing—beneficial for playful humans but due to increasing carbon-emission harmful to flora and fauna—becomes a matter of severe moral questioning. Second, Meinberg argues that a comprehensive co-existential ethics of sport should have a *macro character*. Not just sports(wo)men but also institutions, systems, the natural environment and non-human life that is involved one way or another with the effects of sport on the environment should be taken into account. Third, Meinberg insists on the *pluralist character* of a co-existential ethics of sport. Not just sport-intrinsic norms (fair play, the principle of equality of competitors, etc.) but also external value-systems should be taken into account. The ethics of sport inevitably has to consider and cooperate with other forms of applied ethics, Meinberg argues. Especially with regard to ecological ethics this is, due to the precarious current state of planet earth, also an ethics of survival.

So the decisive question is: what is an appropriate and well-matched relation between humans and nature? Meinberg pleads for a democratic "plural ecology" instead of radical and coercive ecological straightforwardness: "Plurality is a protection against dogmatism, tyranny and absolutism, for which reason the plurality-commandment itself is morally motivated" (p. 139–140). He ends with his fourth particular: a call for a good-natured, obliging and moderate human sportive selfishness. He calls this a well-willing or 'benevolent anthropocentrism', which uses nature as a playground for sport practices only reluctantly.

> *Homo sportivus* should—and because it is more honest this is the real benevolent anthropocentrism—therefore protect natural spaces, truly cultivate them, to protect himself. Yet a co-existential ethics of sport cannot decide if such an insight also will be followed by appropriate acts. Who knows how hard it is for people to convert thinking into moral acting will most certainly not glow with optimism (p. 151).

Meinberg's attempt to design an up-to-date co-existential ethics of sport may look circular. He seems to be aware of this. Therefore, he makes an anthropological turn. For him, a human is a *homo compensator,* a being opposed to nature, but at the same time embodied in and dependent on the natural world. So in this respect humans are rather participants *in* than sheer masters *of* nature. Following the Spinozistic scheme, Meinberg's before-mentioned theses of 'co-existence' and well-willing anthropocentrism implicitly implies the *naturans*-modus. In his view there is no opposition between (unnatural) sport and (natural) nature; they are interdependent. Even more, the concept of 'nature' is anthropomorphic by nature. Nature it is always operational in a prefabricated cultural context.

Consenting with Meinberg's idea of the inescapable benevolent anthropomorphism when it comes to caring for the environment, I contend that what we refer to as created, thus 'natured' in the Spinozistic sense, in every day speech is also always creative, or *naturans*. The creative act of 'seeing' as such constructs the world as a vivid place, not the other way round. It is the eye of the beholder that constructs what is referred to as 'reality'. *Esse est percipi*, to be is to be perceived, after the dictum of George Berkeley (1685–1753), the Irish philosopher and bishop whose work can be characterised as 'subjective idealism'.

Berkeley argued, for instance, that a tree which falls over in a deserted forest properly speaking does not make sound. Although one may reasonably argue that the sound of the falling tree is also there without the ear of the receptive human, the meta-empirical phenomenon 'sound' manifests itself only through the anthropomorphic act of hearing and transforming pulses into consciously perceived 'things'. There is only sound in the ear or mind of the human beholder. Or, in our case, nature only exists by virtue of perception.

> For as to what is said of the absolute Existence of unthinking Things without any relation to their being perceived, that seems perfectly unintelligible. Their *Esse* is *Percipi*, nor is it possible they should have any Existence, out of the Minds or thinking Things which perceive them (Berkeley 1710/1734/2012, p. 13).

This cognitive solipsism however doesn't discharge humans from responsibility, I argue. Once there, the awareness will emerge in us that the perceived world is complex, abhorrent, indifferent, precarious and beautiful at the same time. While all these emotions probably will appear in a different mix in each 'percipient', these are still communicable and discussable. A thread in this ongoing process of exchanging impressions is that the human appearance on earthly scene comes with responsibility—be it for an illusionary world that only exists in the individual soul or mind or a real existing one. A truly sensitive empirical idealism therefore will acknowledge the world as an *organon*, a living and naturing tool that because of its fragility demands loving care.

The paradoxical and problematical relation between what is usually referred to as 'the objective world out there' and human subjects, who unavoidably always interpret and colour the world, is one of the philosophical themes *par excellence*—be it in a sensitive empirical idealistic vein or any other. To flesh out the tension between things and words or ideas (always lingering under the surface), I prefer to oscillate between the practical and the ideal, the perceived and the perceiver, adaptation and mitigation, word and world. Living a worthy and inspired, but nevertheless full and hearty sporting life with as little environmental pressure as possible: this is the heart of the matter.

2.3 Ecosophy of Sport

The first stop in the quest for a more robust, practical and practicable ecological philosophy of sport, that ideally will lead to a more profound change of the way people live their lives, is 'ecosophy'. This portmanteau of ecology, the scientific analysis and study of interactions among organisms and their environment, and philosophy, literally: the love of wisdom, has been coined by Norwegian Arne Naess (1912–2009), the founding father of the deep ecological movement in the early 1970s. Deep and shallow ecology are two key notions in Naess's 'theory'. Shallow ecology refers to all human strivings to create a more sustainable world, say energy neutral Olympics. These practical efforts similes the light green approaches as distinguished by Jarvie & Thornton. Deep ecology, on the other hand, denotes a truly ecosophical dark green mindset to actually change the way humans reel and deal with 'nature'.

In consonance with the idea that practical philosophy is about overcoming the self-imposed limits of meta-ethics and trying to awaken philosophy from its "metaphysical slumber" and its "conceitedness, marginality and isolation" (Zwart 1999, my translation), I argue that the specific situatedness or being-in-the-world of philosophers does play a decisive role in their philosophies.[15] Meaning cannot simply be distilled out of thin air. Real love of wisdom is always in need of naturing surroundings. If not, it fades away in a tasteless metaphysical *aporia*.

The importance of an ongoing dynamic contextual embedding also goes for Naess's holistic ponderings on a better, more co-existent world. His ecosophy is an amalgam of a sometimes vague speculative vitalism (Loland 1996, p. 71) with spirituality, straightforward moralism (Minimise your ecological footprint! Cherish life in its entirety! Enjoy nature properly!) and the linguistic preciseness and nearly mathematical reasoning of analytic philosophy. It is a tactile mixture of personal intuitions and a rather rigid system of hypotheses and prescriptive norms that cannot be understood without reference to the natural habitat from which it arises.

As for Naess's specific biotic roots: he was a noted mountaineer who in 1950 led the expedition that for the first time in history scaled the Tirich Mir (7,708 m), the highest mountain of the Hindu Kush range. He shares his attachment to the remote highland with Friedrich Nietzsche, who in *Ecce Homo* already proclaimed that philosophy is a voluntary and solitary walk through icy mountains,[16] and with Martin Heidegger, who preferred the simple life of thinking, writing, walking and skiing in the Black Forest above the vibrant but hectic and artificial culture of university life.

---

[15] "From word to world, from deduction to induction" (Zwart 1999, my translation).

[16] "He who knows how to breathe the air of my writings knows that it is an air of the heights, a bracing air. One must be made for it, otherwise the danger is no small one of catching cold in it. The ice is near, the loneliness is tremendous—but how peacefully all things lie in the light! How freely one breathes! How much one feels beneath oneself! Philosophy, as I have understood and lived it hitherto, is the voluntary living among ice and high mountains—the seeking-out of all things curious and questionable in existence, everything that has been put under a ban by morality hitherto" (Nietzsche 2004, p. 8).

Only in solitude one can find life in its naked state. Nietzsche, who was always suffering from mood swings towards the negative, felt rather comfortable at the little Swiss mountain village of Sils-Maria, close to the Maloja-pass, an area of "barren ruggedness—rather than prettiness" (Linley, 2014). While Heidegger found his peace of mind at his little den at Todtnauberg, a small mountain village at 1,000 m above sea-level, 20 km south of his *alma mater* in Freiburg, far from the bustle and the overcrowding. There he wrote.

Heidegger cultivates his garden at his den at Todtnauberg, near Freiburg, Germany. (Source: Kinkats & Co)

View from Heidegger's vacation chalet in Todtnauberg. Heidegger wrote most of *Being and Time* (1927) there. (Source: Wikipedia)

Also Naess cherished his refuge. He frequented his Tvergastein hut in the Hallingskarvet massif in order to ponder over the deeper matters of life. This off the grid safe haven plays an important role in Naess 'Ecosophy T'. 'T' is said to represent Tvergastein in this all-inclusive environmental 'theory', which is also a prescriptive guide for leading the good life.

> By an ecosophy I mean a philosophy of ecological harmony or equilibrium. A philosophy as a kind of sofia (or) wisdom, is openly normative, it contains both norms, rules, postulates, value priority announcements and hypotheses concerning the state of affairs in our universe. Wisdom is policy wisdom, prescription, not only scientific description and prediction. The details of an ecosophy will show many variations due to significant differences concerning not only the 'facts' of pollution, resources, population, etc. but also value priorities (Naess cited in Drengson and Inoue 1995, p. 8).

Naess gets an intercultural visit at Tvergastein. (Source: Elementa)

Although Naess's work is permeated with *friluftsliv*—the Norwegian preference for unpolished outdoor activities such as hiking,[17] climbing and kayaking—, his connection with canonized sport is at first sight anything but obvious. At the beginning of his *Outline of an Ecosophy of Sport* (1996), Sigmund Loland stipulates that in the

[17] In 2014 I experienced that quite dangerous, un-secured hikes over windy cliffs, which probably would lead to a nation-wide political discussion in The Netherlands, in Norway are labelled as family-friendly pastimes.

ecological movement "to a certain extent, competition has been seen as the very counter-principle to ecologically sound practice in which cooperation and symbiosis are considered key values" (p. 70). Based on the work of his fellow-countryman Loland, nonetheless, argues that sport, at least to some extent, can be ecosophically justified. He characterizes Naess's ecosophical theory as "a speculative, at times vague vitalism with the clear-cut rationality of analytic philosophy" (p. 71–72). The analytical, puzzle-solving fundamentals of Naess's thinking can be summarized as follows. Ecosophy T consists of a set of norms (N) and hypotheses (H). N1 is the "top norm", which refers to Self-realization! All other norms and hypotheses are logical derivations of N1.

> The hypotheses provide a supporting network, as they are verbal articulations of the relationship between norms. The use of both hypotheses and norms is considered by Naess to be important. Disagreements on norms are often caused by disagreements on facts. By including hypotheses, the chance for locating disagreement increases while at the same time underlining the status of the norms not as absolute and rigid rules, but as prima facie guidelines (p. 72).

The constitutive norms of Naess's ecosophy T can be captured in a few words or sentences ending with an exclamation mark. This renders Ecosophy T into what can be called a holistic ontological system, which consists of a total world view, with a definite sender (Naess) but with no definite receivers, Loland argues. "The norms are considered valid for all life forms in all situations. Naess's ecosophy is in this way a cosmology, a total view of how 'all things are interconnected'" (p. 72).

Loland stipulates that Naess is reluctant when it comes to what he calls 'precisations' of the relation between his 'top norms' and their realization in T(heory) terms, "as general and vague terms that represent powerful intuitive ideas that are shared by many and inspire a variety of inspirations" (p. 72). It seems as if true ecosophy only can be 'felt' by true adepts. Nevertheless, on the prescriptive, straightforwardly normative side Naess distinguishes three ways of interpreting N1: *Self-realization!*

The first theory that people may adhere to (T1) is *ego-realization*. This represents the traditional view in Western culture, with "a predominant anthropocentric, individualist, and utilitarian world view … in terms of ego-realization, the maximization of narrow self-interest" (p. 72). Nature is instrumental in this view. T2 refers to *self-realization* in a somewhat wider sense, since in this perspective also other humans are included. "It can, perhaps, be linked to conceptions of self realization within mainstream humanistic psychology and moral philosophy—for example, to traditional interpretations of *eudaimonia* or 'the good life[18]'"(p. 72).

> Naess prefers, of course, a less specialized precisation—T3. T3 refers to an expanded, ecological Self that is realized in a process of deep identification with individuals of all life forms. This idea is linked to holistic traditions of both Western and Oriental thought, to notions such as 'the universal self,' 'the absolute,' *atman* (sanskrit), and to ideas of the comprehensive structure of the self found in the writing of thinkers such as Spinoza and Gandhi (p. 73).

---

[18]The good life in the deeper Aristotelian sense, beyond shallow pleasure or hedonism. *Eudaimonism* often has a 'mental' connotation. In due course I will argue for a more 'physical' interpretation of well-spiritedness.

Building on T3, the magnanimous form of the *'ecological Self!'*, Loland outlines his argument for a practicable conception of sport that, despite its competitive (so to some extent also always destructive) disposition still can stand the ecological litmus-test. Key in this effort is to radically change[19] the way humans deal with sport—and thus inevitably also with their natural habitat, so with life in general—are a constant hovering between holistic *'Self-realization!'* on the one hand, and the idea of 'biospheric egalitarianism' or 'the democracy of all life forms' on the other.

Naess overcomes the classical Cartesian subject-object-divide by means of his 'relational field ontology', in which humans are not interpreted as superior to other species, but as humble participants in the larger scheme of being. "Objects, plants, animals, persons, or practices like sports cannot be fully understood as single entities and in isolation, but only as relational wholes" (p. 73). All life forms possess intrinsic value and thus have "a universal right to live and blossom" (p. 74).

Beyond Meinberg's before-mentioned well-willing anthropocentricism, which remains in the realm of classical metaphysics, Naess argues for replacing classic subject-object dichotomies (that put humans at the top of the evolutionary pyramid)[20] by a more inclusive theory. He does so by proposing 'relational field ontology', a comprehensive and inclusive theory of being that envisions humans as co-evolutionary parts of a whole rather than (just) benevolent stewards of nature. Entities or things, including humans, are knots or junctions within the holistic field of being. To be is to be perceived.[21]

> All statements 'about the thing' are relational statements: statements like 'thing A is B' are in Ecosophy T abandoned in favour of 'thing A is B in relation to C' or the relational thing AC has the quality B (Naess, cited in Loland 1996, p. 73).

In Naess's relational field ontology the distinction between humans and environment becomes artificial and meaningless, thus unnecessary. It is the interrelation between entities or things, formerly known as subjects and objects that counts. Still, it cannot be denied that humans have a rather unique position. They have a special capability of putting the interconnection between all there is into meaningful words and thoughts. This ability to verbalize and literally re-flect hopefully will lead to acknowledging that we should strive for peaceful coexistence. *Eros* is the widest interpretation of the identification of humans with other species. "In love one loses

[19] Cfr. Keulartz (2005) for a more detailed account of mitigation (a radical change of lifestyle, making a u-turn) versus adaptation (adjusting your lifestyle, steering away) and for how only boundary work and post-metaphysical pragmatism can help us out of the stalemate between a light-green market-liberalism and radical dark green approaches. In Chap. 5 *Continental Pragmatism: Enduring Life in the Strenuous Mood* I will propose a synthesis of both visions.

[20] Mark Twain (1835–1910) already questioned the supposed supremacy of mankind: "Man has been here 32,000 years. That it took a hundred million years to prepare the world for him is proof that that is what it was done for. I suppose it is, I dunno. If the Eiffel Tower were now representing the world's age, the skin of paint on the pinnacle-knob at its summit would represent man's share of that age; and anybody would perceive that the skin was what the tower was built for. I reckon they would, I dunno" (1993, p. 106).

[21] In Chap. 6 *Continental Pragmatism: Enduring Life in the Strenuous Mood* will argue that Naess's relational field ontology overlaps with William James's subtle pragmatism.

part of one's identity by gaining a greater identity, something that in its truest sense cannot be spoken of" (Naess 1989, p. 11).

Naess's universal love principle[22] provides a course of action worth striving for, Loland contends. On the one hand humans have to become fully aware of their 'greatest good', their ultimate *telos*: to strive for maximized Self-relation! in consonance with non-human species.[23] On the other hand humans have to realize that they are unique because of their lack of biological specialization.[24] This peculiarity enables them "to adapt to situations in almost all parts of the ecosphere" (p. 74). Because of their *Sonderstellung* ('special position' (Max Scheler 1928/1975)), their exceptionality, humans are more powerful than other species. They are capable of reconsidering their brute supremacy over other life forms. They possess the power of reason. They are able to ponder over their own actions on earth, "and in fact influence Self-realization through deliberate action" (p. 74). For Naess this implies an ethical responsibility in the way we relate to 'others', since we can "consciously conceive the urge other living beings have for self-realisation" (Naess, cited in Loland 1996, p. 74).

Loland goes on to say that Naess's call for human responsibility is a matter of inner conviction, not of external duty.[25] "This is the true realm of freedom in a Spinozistic sense, 'a way home' to what we really are. Naess emphasizes the importance of moving from ethics to ontology and back" (p. 74). The ongoing oscillation between is and ought paves the way for introducing so-called "Self-realization

---

[22] Once more Ortega y Gasset prima vista seems to have the oldest rights in this case (cfr. Footnote 18). Ilundáin-Agurruza (2014, p. 286) argues that, next to 'circumstance' and 'salvation', 'love' is the third pillar of his ratio-vitalism, which paves the way for an optimistic concept about sport. "Phenomenologically, he uses sport to reveal the structure of life and how we experience it" (p. 288).

Learning to love something or someone is about making the other the centre of our life. In keeping with thick holism, I conceive this as a perspectival shift that entails the union of lover and loved. It is about unity, about bringing together what was asunder before. We cannot live without the loved object—we consider it part of ourselves—and through concatenation of those things about which the loved one cares and so on, we ultimately are led into a Spinozistic 'pan-holism' where all is united. Ortega's and Spinoza's holism correlate, for different reasons, with Eastern views that endorse egolessness, emotionally and ontologically. This bypasses the perennial problem of the one and the many and the subject/object dichotomy, if for a time: 'Love […] binds us to things, even if only temporarily' (Ortega y Gasset 1963, p. 33). One who acts thus becomes a noble exemplar for Ortega and is willing to take risks most others skirt around (Ilundáin-Agurruza 2014, p. 295).

I argue, however, that Naess's interpretation of the spinozistic *unio mystica* is far more receptive to contemporary environmental issues. Whereas Ortega largely stays in the realm spiritual connection, Naess also includes pets and plants.

[23] It should be noted, however, that Naess makes an exception for the fulfilment of so-called 'human basic needs'. The problem is, however, that these seem to change over time. These days many (Western) people will consider a yearly air trip to a sunny destination a basic need.

[24] According to Arnold Gehlen (1940/2009), from a purely biological perspective humans are *'Mängelwesen'*, deficient beings.

[25] Further on I will argue that from a pragmatic point of view it doesn't matter if acts are undertaken out of inner or conviction or outer obligation. It is the result that counts.

potentials" (p. 75), as Naess calls the potency of different forms of life to bring their capacities into actuality. Self-realization potentials result in adding something to the vast whole of life per se. No matter how futile the concerned 'individual', actualised potentiality always results in an increase of ecosophical joy. Say a flower that attracts bees by blossoming or just pleasures the human eye. "The key ideal of Ecosophy T, then, is this: Unity and diversity of life" (p. 75). Loland summarizes Naess's ideal ecosophical scenario as follows:

> N 1: Self-realization!
> H1: The higher Self-realization attained by anyone, the broader and deeper the identification with others.
> H2: The higher Self-realization attained by anyone, the more its further increase depends upon the Self-realization of others.
> H3: Complete Self-realization of anyone depends on that of all.
> N2: Self-realization for all living beings!
> H4: Diversity of life increases Self-realization potentials.
> N3: Diversity of life!
> H5: Complexity of life increases Self-potentials.
> N4: Complexity!
> H6: Life resources of the Earth are limited.
> H7: Symbiosis maximizes Self-realization potentials under conditions of limited resources.
> N5: Symbiosis! (p. 76)

However attractive this ideal scheme for a symbiotic ecosophy may seem, Loland also points at potentially controversial consequences of the deep identification with the Self-realization potentials of all creatures, from upright walking mammals with an oversized neo-cortex to the smallest of insects.[26]

> For example, we might, under certain circumstances, reach the conclusion that human beings ought to recommend their own withdrawal as the dominant life form on earth[27] to promote other life forms to live and blossom; this withdrawal may contribute more to the Self-realization for all. Such norms, critics could argue, cannot serve as a common basis for society at large (p. 76).

A brief turn to everyday life illustrates that the withdrawal-stance is not very likely to happen in a man-made world. Due to a plethora of technological innovations we increasingly tend to see nature as malleable; as more or less co-evolutionary with

---

[26] Recent ethological studies (e.g. De Waal 2006/2016) show that the mind and self-consciousness are a matter of evolutionary degree rather than uniquely human. Stephen Jay Gould even has argued (Scientific American, October 1994) that "[o]ur impression that life evolves towards greater complexity is probably only a bias inspired by parochial focus on ourselves." Over time 'the wall of simplicity' (bacteria) has proven to be more adaptive and more resistant to changes such as the meteorite impact on Yucatan, some 65 million years ago, which caused the Triassic-Jurassic extinction. The life-style of bacteria therefore has remained the most common and most successful. "We must understand that little twigs [e.g. *homo sapiens sapiens*, rw] are contingent nubbins, not predictable goals of the massive bush beneath" (p. 91).

[27] This might seem quite appropriate considering the current state of affairs (sea-level rising due to carbon emission, large scale deforestation, human-induced extinction of species that exceeds the Triassic-Jurassic extinction by far, over-population etc.) But the pivotal question than is: who is in charge of the human plant planet earth meanwhile has become? Who will control the control-panel? There seems to be no way back, thus ludditism will not help, eventually.

our irrepressible desire to conquer planet earth. We have taken over control of planet earth in a very un-ecosophical, consumptive and speciecist way. Rather than staying at home and close to our little den, our personal Walden, we try to develop cleaner ways of long distance travelling. Rather than restrain ourselves we employ the *contradictio in adjecto* of 'sustainable growth'. And, eventually, rather than caring for other biological species we exterminate them.

The persistent habit of solving problems by un-problematically applying external τέχνη (*techné,* art, craftsmanship) instead of exploring the internal potentiality of φύσις (*phúsis*, nature)[28] seems very hard to break, all in all. Even more, contemporary elite 'record sports' seem to confirm the logic of unbridled technological progress (Loland 2001). The hunt for records equals the adage of un-satiability: enough is never enough. Records need to be broken, if necessary by thousandths of seconds. At a closer look, this insatiability, this ruthless hunt is the ultimate example of the law of diminishing returns. Loland stipulates that because of this unrelenting attitude of progress at all costs Naess is sceptical towards high end competitive sport, "with its focus on specialization, achievement, and standardization of play areas and play actions" (p. 77). The fixation on quantification, the habit of reducing quality and beauty to sheer numbers, makes elite sport prone to shallow and instant hedonist satisfaction, rather than enjoying the deeper delights of true eudaimonism.

Nevertheless, Loland still thinks that experiencing "what we may call ecosophical joy" in sport is possible, since in sport there are also many "paradigmatic examples of autotelic activities in which people take part due to values realized in the activities themselves" (p. 77). Building on Naess's ideal ecosophical scenario he therefore proposes a set of hypotheses and norms for ecosophically sound sport practices. To avoid confusion with Naess's typology he uses small letters to concretize N1: Self-realization! for sportive ends, which results in postulating h(ypothesis) 1: "Self-realization in sport depends on the realization of joy" (p. 77), and n(orm)1: "Joy in sport!" (p. 77). Loland then argues that:

> the kinesthetic joy of well-exuceted motor actions can be just as strong for the skiing novice who for the first time feels the rhythm of a good turn as for the expert who after long and tedious training is able to include a new detail into a larger technical repertoire (p. 79).

[28] I here refer to Aristotle's idea that *phúsis* is its own source of bringing its potentiality to actuality, which perfectly fits in Naess's ecosophical scheme. *Techné,*on the other hand, requires a source of motion outside itself, thus is at best shallow ecological. This resembles Sloterdijk's distinction between nature-friendly homeo-technology and alienating allo-technology already referred to in the introductory Chap. 1 *Prologue: The Good Life, Asceticism and Sustainable Cycling.* Vincent Blok (2014) points at the more intricate relation between *phúsis* and *techné* in Heidegger. "As a representation of the *phúsis*, *techné* has to be understood as an *addition* or *supplement* to the original *phúsis*. Why? Because *phúsis* has the tendency to conceal itself, only a technical supplement or re-presentation of the *phúsis* is able to give us access to the original *phúsis*. It is in this respect, that we can say that the *techné* is demanded by *phúsis* in order that it can reveal itself" (p. 324). A Heidegger-inspired take on (sustainable endurance) sport will be developed in Chap. 5 *Ascetic Practices, Hermeneutical Cycles and Ecosophical Endurance.*

When sport becomes boring, about repeating the same and dull routine over and over again however, Joy in sport! is over, since there only remains width instead of depth. Loland further formalizes his sport philosophical argument as follows:

> h2: Joy in sport depends upon exploration of the unity and diversity of embodiment.
> h3: Exploration of the unity and diversity of embodiment depends upon skill development in an optimal balance between width and depth.
> n2: Skill development in an optimal balance between width and depth! (p. 80)

The resulting ecosophical imperative finally enables Loland to formulate a practical rule of thumb for morally measuring sporting acts that people intend to undertake.

> Naess's slogan of 'simplicity in means, richness in ends' is true in sport as in other areas of life. At the same time, technology can add to the ecosophical joy of sport and thus serve the Self-realization process. In sport, we can establish meaningful relations to technology, and technology can expand our experience of the unity and diversity of life. If increase in joy is able to outweigh on a long term-basis the ecological costs of production and application, sport technology can be ecologically justified (p. 84).

Applying the ecosophical joy calculus—rich sporting ends enabled by technological means as simple as possible—to the admissibility of new technology in specific sport practices means a full 'yes' to the introduction of the carving ski and telemark technology, since these facilitating techniques according to Loland increased the pleasure of skiing significantly.

Sigmund Loland, a joyful skier with a deep philosophical look. (Source: Idrottsforum)

Applying the ecosophical fairness principle as an ethical variant of Occam's ontological razor would imply a full 'no' however to the introduction of aerodynamically integrated 'invisible' bicycle brakes in time trials and triathlon, to switch over to a sport practice that isn't mentioned by Loland in his article.[29] Since these high

[29] Sport philosophers often carry their own background into the debate. Loland's roots are in skiing. Mine are in endurance sport, especially long distance cycling. Skilfulness, agility and exhilarating joy versus perseverance, resilience and diligence. Loland refers to Spinoza's concept of *hilaritas*

end expensive technological novelties makes the top-level riders just a few seconds faster over 40 km in a non-drafting time trial. This is an exemplary case of diminishing returns that do not outweigh long-term ecological costs, thus do neither lead to increased ecosophical joy nor to Self-realization!.

Loland admits that the most controversial norm in his subtle effort to develop a comprehensive ecosophical framework with an appropriate set of hypotheses and norms is the demand to practice 'sport in closeness to nature!' This back-to-nature principle implies that sports that have been practiced indoors for decades, like handball, basketball, etc., would have to change dramatically and regress to a primordial state where bumpy outdoor pitches and weather conditions are highly influential on the game. But also sports which are still performed 'in the wild'—from skiing and paragliding to freestyle rock-climbing and trail-running—often are at odds with true ecosophy. These nature sports, as already argued, often have a detrimental effect on ecologically vulnerable biotopes, by the practices themselves, but also by the facilities that are built to enable them and the travelling to and fro the places of sportive interest. This invasiveness pushes the ecosophical balance towards the negative.

Especially contemporary winter sports are prone to the inner tension between a sustainable imago and concrete ecological impact. This turned out to be the case with the Vancouver Winter Olympics, which were intended to be dark green but were actually light green. Technological developments usually result in the paradox of bringing about naturalness by unnatural means. Loland now points at winter sports from a different angle: "Alpine and telemark skiers demand energy consuming ski lifts and machine-prepared snow" (p. 85). While techno-innovations at the individual level have increased the possibility for experiencing sport-ecosophical joy, overall they do not meet the norm that sport technology can be accepted when the strengthening of joy outweighs the ecological costs. What at a glance looks like experiencing bare and unconcealed life in pristine 'natural' surroundings, unveils itself as nothing but anthropocentric artificiality: the ultimate urbanisation of the sublime (Zwart 2017, p. 13). Although skiing probably will lead to self-knowledge in the sense of Naess's ego-realization, there is no Self-realization! that lets all life forms blossom. Skiing in this respect thus is 'un-ecosophical'. Zwart discerns a tendency "to disavow the ecological and technological dimensions of skiing, so that the nature experiences evoked are much less natural than they purport to be, while environmental awareness is silenced rather than promoted" (p. 13).

It is justified to conclude that Loland's proposed outline for an ecosophy of sport offers an escape route signposted by viable norms for deciding whether or not to act in a specific situation. When relatively simple means result in quite rich ends one should at least consider the admissibility of a specific sport practice. As already

("a strong positive emotion with thestructure of a holistic Gestalt" (1996, p. 77). I rather feel sympathy for Peter Sloterdijk's (quasi-)Sysiphean idea of man the diligent practitioner. Loland doesn't deny that "periods of monotonous, hard work are sometimes necessary to reach deeper insight and values"(p. 80)—and thus in the end may result in augmented ecosophical joy—, for Sloterdijk the 'joy' is in the hard work itself. This different basic attitude towards repetitive practices will be tested in the following chapter.

stipulated, problems arise however when new technologies, such as the introduction of carving skis, at the individual level indeed open up opportunities for enjoying skiing significantly, but at the collective level result in substantial environmental damage. What to do when under the veil of nature experiences, artificial snow is produced and the Alps are run-down? How to find the right mean between our universal responsibilities for maximized Self-realization! for all species and individual self-knowlegde? Can sport fit in the Spinozistic way home to what we really are? How to determine the right mean between deep ecological ponderings and shallow but practical and practicable sustainability? What type of sport practiced by whom are we discussing anyway?

2.4 Taking the Measure of Record Sports

In *Record Sports: An Ecological Reconstruction* (2001) Loland further specifies his ecosophical attempts by distinguishing three categories of sports. First, he distinguishes 'record sports'. These are sports where exact measurements of mathematical-physical entities within a standardized spatio-temporal framework are possible, such as swimming, athletics and weightlifting. Second, there is 'quasi-record sports', such as marathon running. Sometimes people talk of records here, he reasons, but this is inaccurate since conditions (e.g. weather) and trails (e.g. flat or mountainous) differ from race to race.[30] Loland's third category consists of 'games',[31] such as football, basketball, baseball, etc. These are non-record, but nevertheless highly competitive sports. In this sense they also raise severe ecosophical concerns, from aggression on the pitch—which definitely is at odds with the ecosophical plea for peaceful co-existence and the call for the flourishing of all life forms—, to the environmental impact of football, basketball and baseball. A fourth category, which isn't mentioned by Loland, that can be discerned are so-called 'juried sports', such as gymnastics, figure-skating and (horse) dressage, whilst also sports such as boxing and martial arts to some extent are juried.

Since they are popular, practiced all over the world and highly visible in the media, thus exemplary, Loland concentrates on 'record sports', in particular modern Olympic athletics.[32] After a brief review he puts forward his challenging 'record sport dilemma', which replaces the pre-modern conviction that *Enough is enough!* by the modern logic of *Enough is never enough!* Every new record has to be broken,

[30] Also the mentioned relatively slow Norseman triathlon and the relavitely fast Roth triathlon are quasi-record races.

[31] In Chap. 5 *Ascetic Practices, Hermeneutical Cycles and Ecosophical Endurance* I will argue that the philosophy of sport so far has overestimated the un-committal game-character of sport and underestimated endurance sports, which usually are not very 'ludic'.

[32] It should be noted that in the ancient Olympics included running, long jump, shot put, javelin, boxing, *pankration* (a form of martial art, combining wrestling and boxing) and equestrian events. So there were no ball (or puck) games, nor gymnastic disciplines!

over and over again, by tenths, hundredths or recently even thousands of a second. The official motto of the Olympics *citius, altius, fortius*—ever faster, higher, and stronger—is a perfect as well as a perfidious expression of this literally unlimited striving for progress.

There obviously is a downside to the record as the 'eternal axiom of sport', to use the words of Pierre de Frédy, baron de Coubertin (1863–1937), founding father of the modern Olympics. Although (economic) progress often is considered as *conditio sine qua non* in our times, we cannot go on forever with exploiting the limited resources of our planet. Loland: "The continuous quest for new records is built on the impossible quest for unlimited growth in limited systems" (2001, p. 130). He argues that the tendency to replace nature's "curved lines and unpredictable diversity" by "the straight, geometrical coordinate system of the arena" is a socio-cultural dead end, however.

> The ideal of record sports portrays a one-dimensional social logic and a mono-culture version of sport. All athletes strive for the same objective goal of record setting, and they prepare for and specialize in exactly the same kind of performance over and over again. Hence record sports have the characteristics of a non-sustainable system (p. 133).

Five years later again Loland publishes a next ecosophically oriented article, entitled *Olympic Sport and the Ideal of Sustainable Development* (2006). Here he reflects on the ambition of the IOC "to engage in protection of the environment" (p. 144), already discussed at the beginning of this chapter. He reconsiders his record dilemma and connects it to environmental issues in Olympic sport, especially record sports, and specifically sprinting, once again. For good reasons, since the 100-metre dash probably is the most prestigious Olympic discipline. Loland argues that the sport-record that has to be broken over and over again in the end is untenable from an ecosophical point of view.

> On the 100 metre sprint, a race on 10 seconds requires runners that run below 10 seconds in the current and next generation of sprinters. From the record perspective, performances above 10 seconds are used once and for all. Hence, every new record represents the use of a nonrenewable performance resource in the sport to which it belongs. Every record means a missed opportunity for performance for other current and future athletes. Record sports are non-sustainable in that they require unlimited growth in limited systems … If Olympic sporting is life is to continue indefinitely, record sports have to be either abandoned or reformed (p. 147).

Loland therefore advises the IOC to drop the idea of measuring performance in absolute arithmetical units. His alternative for the numerical obsession with records is an ordinal ranking for competitors according to the order in which the cross the finishing line or, if sprint competitions are made up by a series of competitions, by giving points such as 5 for a win, 3 for a second place and so on. But this alternative ranking is not enough to achieve ecosophical justice, Loland contends:

> Requirements on standardization of record events can be transformed into requirements on equal conditions in a variety of ways. For instance, competition schemes could be varied by introducing a variety of distances (sprinters could run from between 30 and 300 metres, but never the exact same distance twice) and surfaces (sand, running tracks, turns, slightly uphill, slightly downhill etc.) … In such a reformed sprint-running scheme, a larger variety of sprint talents can blossom (p. 150).

One could question, however: what are the consequences of this ecosophically revised sprinting scheme for real ecology out there in the remains of what we still consider nature? How do diversified running tracks, varying from sprinting some 300 m on permafrost in the northern tip of Sweden and 150 m in the Gobi-desert to 30 m on high altitude rock-solid ground in Kigen, Kenya contribute to a more ecosophical conception of sport? Even more, if these varying tracks and distances have to be imitated during an Olympic tournament: what are the costs and how much means would this diversified ecosophical sprinting celebration actually use up?

We may discern an analogy between the blossoming of various sprint talents and the blossoming of biological species. More variety usually results in more resilient ecosystems. One can also ponder on the special benefits of an ecosophized sprint circus for a better intercultural understanding of people all over the world. However, history teaches us that we probably will forever be longing for athletes that challenge and push the limits in terms of hundreds or even thousands of a second. It is hard to imagine the Olympics, with its tradition of a strong focus on quantitave athletics, changing in the ecosophical direction Loland suggests.

Even more, a world record-breaking 100 m sprint race can be ecologically justified. Imagine for instance a sprinter that converts himself into a vegan hermit, a true ascetic, a shining example for mankind who keeps on sprinting, never leaves his home country nor owns a car, but is still able to set a 9,44 s[33] or so world record on a rickety, outdated local athletic track. What's wrong with this unlimited growth in a definitely more sustainable, less resources consuming system? A group of athletes travelling all over the world to visit the ecosophical, intercultural and multi-(sur) faceted sprint series which requires a series of diversified sprinting skills, Loland suggests, will probably do more harm to the environment.

Sympathising with William James's pragmatic rule of thumb that truth is to be found in consequences of concrete acts rather than in (good) intentions, my advice to the IOC would be quite the opposite of Loland's call to concentrate first and foremost on our deep ecological thoughts with regard to our (sportive) being in the world. Redefine your institutions, and minimize the ecological footprint of the games within the current system. Then switch over to sensitising concrete sport practices and individual sports(wo)men towards more 'Self-realization!' through symbiosis with 'nature' in the naturing sense. Nature as homeo-technology that largely follows the never fully predictable curves of *phúsis* rather than ironing them out.

In order to further substantiate my claims, in the following Chapter I will attempt to answer the three decisive ecosophical questions Loland ends his *Outline of an Ecosophy of Sport* with. I will do so by particularly concentrating on asceticism and endurance sport practices.

---

[33] Mathematician Reza Noubary calculated this as the ultimate time for the 100 m dash. After Usain Bolt's astonishing world record of 9.58 in Beijing 2009 he however proclaimed that his prediction probably would go down a bit (Yong 2012). Supposed ultimate records always have been proved human constructs. Breaking the 10 s barrier at the 100 m sprint was thought to be impossible for a long time and also Roger Bannister first sub 4 minute mile (1954) had been thought an impossible hurdle to overcome for a long time. One never exactly can tell how training techniques will develop and evolution goes. Of course there is a logical absolute limit: athletes will never be able to arrive earlier at the finish-line than they start. But on the other hand there is the possibility of breaking records by ever smaller units, such as ten thousands of seconds.

References

Anker, P. 1998. On ultimate norms in ecosophy T. *The Trumpeter – Journal of Ecosophy* 15 (1): 1–4 http://trumpeter.athabascau.ca/index.php/trumpet/article/view/152/1323. Accessed 9 Sept 2017.

Berkeley, G. 1710–1732/2012. *A treatise concerning the principles of human knowledge*. London (Ed. by D.R. Wilkins. Dublin).

Bloemhof, L. 2017. De Volkskrant, June 17 2017, "Stukje voor stukje vermoorden we onze planeet" (Interview with Lewis Hamilton). June 17. http://www.volkskrant.nl/magazine/-stukje-voor-stukje-vermoorden-we-onze-planeet~a4501155/ Accessed 28 Oct 2017.

Blok, V. 2014. Reconnecting with nature in the age of technology. The Heidegger and radical environmentalism debate revisited. *Environmental Philosophy* 11 (2): 307–332.

Borken-Kleefeld, J., T. Bernsten, and J. Fuglestved. 2010. Specific climate impact of passenger and freight transport. *Environmental Science and Technology* 44 (15): 5700–5706 http://pubs.acs.org/doi/full/10.1021/es9039693.

Dawkins, R. 1987. *The blind watchmaker*. New York/London: W.W. Norton & Company.

De Waal, F. 2006/2016. *Primates and philosophers: How morality evolved*. Princeton: Princeton University Press.

Drengson, A., and Y. Inoue, eds. 1995. *The deep ecology movement: An introductory anthology*. Berkeley: North Atlantic Publishers.

Galton, F. 1886. *Regression towards mediocrity in hereditary stature*. Anthropological miscellanea: 246–263. http://galton.org/essays/1880-1889/galton-1886-jaigi-regression-stature.pdf. Acessed 26 Sept 2017.

———. 1892. *Hereditary genius. An inquiry into its laws and consequences*. London: Macmillan & co http://galton.org/books/hereditary-genius/text/pdf/galton-1869-genius-v3.pdf. Acessed 26 Sept 2017.

Gehlen, A. 1940/2009. *Der Mensch. Seine Natur und seine Stellung in der Welt*. Wiebelsheim: Aula.

Gould, S.J. 1994. The evolution of life on earth. *Scientific American*. October: 85–91. http://evolution.binghamton.edu/evos/wp-content/uploads/2009/08/Gould_Evolution-of-Life.pdf . Accessed 24 Oct 2017.

Heidegger, M. 1927/2008. *Being and time*. New York: Harper & Row.

Ilundáin-Agurruza, J. 2014. José Ortega y Gasset: Exuberant Steed. *Sport, Ethics and Philosophy* 8 (3): 285–314.

International Olympic Committee. 2012. *Sustainability through sport – Implementing the Olympic movement's Agenda 21*. https://stillmed.olympic.org/Documents/Commissions_PDFfiles/SportAndEnvironment/Sustainability_Through_Sport.pdf . Accessed 9 Sept 2017.

Jakob, E., and G. Liedtke. 2004. Nachhaltige Entwicklungen im Natursport. *F.I.T. Das Wissenschaftsmagazin der Deutschen Sporthochschule Köln* 9 (2): 14–18.

Jarvie, G., and J. Thornton. 2006/2012. *Sport, culture and society – An introduction*. Abingdon: Routledge.

Keulartz, J. 2005. *Werken aan de grens: De spanning tussen diversiteit en duurzaamheid (Inaugural lecture)*. Nijmegen: Thieme media center.

Linley, J. 2014. *Friedrich Nietzsche: A sea of tranquillity and philosophy.* https://www.swissinfo.ch/eng/a-sea-of-tranquillity-and-philosophy/12822 . Accessed 25 September 2017.

Loland, S. 1995. Coubertin's ideology of Olympism from the perspective of the history of ideas. *Olympika: International Journal of Olympic Studies* 4: 49–77.

———. 1996. Outline of an ecosophy of sport. *Journal of the Philosophy of Sport* 23: 70–90.

———. 2001. Record sports: An ecological critique and a reconstruction. *Journal of the Philosophy of Sport* 28: 127–139.

———. 2006. Olympic sport and the ideal of sustainable development. *Journal of the Philosophy of Sport* 33: 144–156.

Meinberg, E. 1991. *Die Moral im Sport, Bausteine einer neuen Sportethik*. Aachen: Meyer & Meyer.

Naess, A. 1989. *Ecology, Community and Life Style* (Trans. and ed. David Rothenberg). Cambridge: Cambridge University Press.

Nietzsche, F. 2004. *Ecce Homo: How one becomes what one is & the antichrist: A curse on Christianity*. New York: Algora Publishing.

Ortega y Gasset, J. 1963. *Meditations on Quixote*. New York: W.W. Norton & Co.

Scheler, M. 1928/1975. *Die Stellung des Menschen im Kosmos*. Bern/München: Francke.

Spinoza, B. 1996. *Ethics*. London: Penguin.

Tamboer, J., and J. Steenbergen. 2000. *Sportfilosofie*. Leende: Damon.

Theroux, P. 1984. *Half moon street*. Boston: Houghton Mifflin Company.

Twain, M. 1993. *What is man?: And other philosophical writings*. Vol. 19. Berkeley: University of California Press.

Van den Born, R. 2007. *Thinking nature: Everyday philosophy of nature in the Netherlands*. Arnhem: Roos & Roos.

Yong, E. 2012. *Will we ever … run 100m in under 9 seconds?* BBC future, July. http://www.bbc.com/future/story/20120712-will-we-ever-run-100m-in-9-secs. Accessed 26 Nov 2017.

Zwart, H. 1999. Praktische filosofie als filosofische praktijk – en als teratologie. *Filosofie & Praktijk* 20 (4): 169–178 http://www.academia.edu/21789188/Praktische_filosofie_als_filosofische_praktijk-en_als_Teratologie. Accessed 27 Sept 2017.

———. 2017. Skiing and its discontents. *Sport, Ethics and Philosophy* 11 (3): 323–338. https://doi.org/10.1080/17511321.2017.1341950.

Chapter 3
Answering Three Ecosophical Questions: Asceticism

I am to develop into a fakir of coexistence with everyone and everything, and reduce my footprint in the environment to the trail of a feather

(Peter Sloterdijk 2009/2013, p. 449).

Abstract Sigmund Loland ends his *Outline of an Ecosophy of Sport* (1996), advanced in the previous chapter, by proposing three pivotal questions that have to be taken into account when it comes to the sport-ecosophical litmus-test.

1. *What are the implications of the norm on ecosophical joy in my specific sport practice?*
2. *How should I, in my sport context, relate to norms for developing skills in width and depth, for playing to win and for applying only ecosophically sound sport technology?*
3. *What can be done to promote sport training and competition in closeness to nature?*

As a practical philosopher and as a philosophical practitioner, I now will try to give provisional answers to these pivotal questions. I will do so from the perspective of (outdoor) endurance sports. To focus my thoughts, I will concentrate on running and particularly cycling, paved and unpaved, from elite athletes to dedicated age-groupers and joggers and weekend warriors.

Endurance sports do not fare well in celebrating the multitude of motor and physical skills that Loland seems to presuppose. They are about diligence, repetition and sticking to a program in order to be able to cover a certain distance or to complete a race. Fostering endurance is a continuum of work in diverse physical states, rather than practicing for a few moments of exhilarating joy. Also Loland acknowledges the benefits of a dedicated training regimen. Longer periods of hard and monotonous work can to some extent be ecosophically justified, Loland reasons, but just as an instrumental means to an end: "Work is acceptable as a means to increased joy and perfection" (p. 80).

R. Welters, *Towards a Sustainable Philosophy of Endurance Sport*,
Library of Ethics and Applied Philosophy 37,
https://doi.org/10.1007/978-3-030-05294-2_3

Although they may seem fairly dull and repetitive indeed, sports that take a substantial effort are increasingly popular with the sedentary, urban and over-stimulated masses in the search for counterbalance, though. A certain physical constitution obviously is advantageous when it comes to endurance sports. Next to possessing basic health, a somewhat slim body may be helpful as a point of departure for fresh(wo)men in running and cycling. However, perseverance and stamina are far more important when it comes to cultivating staying power—and these ascetic virtues are within reach for most of us. Therefore endurance sports can more easily effect a change for the better than skill and agility demanding elite sports.

Because of this high potential for change, I now (in the tough spirit of endurance sport) will try to answer the three key questions Loland proposes as a guideline for ecosophically good sport. To strengthen my critical assessment of Loland's sport-ecosophical blueprint, Peter Sloterdijk's plea for a radical change of our lifestyle by means of a well-understood 'ascetology' will be put in position. If properly performed, this general training theory will result in *metanoia*, a radical personal change of an unsustainable life-style, or at the collective level even in a 'renaissance' of durable virtues.

Keywords Ecosophical questions · Endurance sports · Ascetology · Durable virtues

3.1 Implications of the Ecosophical Norm

First ecosophical question

What are the implications of the norm on ecosophical joy in my specific sport practice?

Loland implicitly argues in favour of an *autotelic* conception of sport: sport as a playful end in itself. In this vein sport is an activity for its own sake, in the sense of Naess's notion of "dwelling in situations of intrinsic value" (cited in Loland 1996, p. 86). However, for many twenty-first century people, sport is not just a non-serious, gratuitous, *ludic,* playful recreational activity or just an appropriate health enhancing technique. For the truly dedicated practitioner sport is rather a truth- or knowledge-seeking lifestyle with an added, supra-playful and supra-health value.

Sport philosopher Heather Reid—in her younger days a promising cyclist, aiming to qualify for the 1984 and 1988 Summer Olympics—is one of the major proponents of the idea that sport is more than just un-committal fun and innocent play. "What my own study of these phenomena reveals is that sport's social and educational benefits derive not from its playful character, but from its philosophical origins as a knowledge-seeking activity" (2009 p. 40). For her, sport is (also) a means to an end. It is as large as life.

> Now, as a middle-aged college professor who never did stand upon that Olympic podium, I can nevertheless say that sport brought me a long way towards being the kind of self I hoped

> would win a medal. Looking back at my early athletic career through the lens of my academic training, I now see the connections to Platonic ideals, Aristotelian virtue-ethics, and Stoic self-creation. I can say I was a philosophical athlete before I understood Plato, or the Greek conceptions of excellence, education, and happiness…. By taking a philosophical approach to sport, athletes of all ages, shapes, and sizes can reclaim the educational value of athletics as it was championed in ancient Greece by such great thinkers as Plato, the wrestler (2002, pp. xii–xiii).

The longing for philosophical athletism may be the reason why so many dedicated non-professional and non-elite practitioners of endurance sport still work out six times a week—or more.

Now, for the sake of the argument, imagine being a dedicated rider and runner rather than an occasional cycling weekend-warrior or a jogger (Cfr. Hochstetler and Hopsicker 2012). Since you are doing fairly, but not extremely well in both disciplines, and because you do not like to ride in a 'peloton', you specialize yourself in the duathlon,[1] or run-bike-run. There a non-drafting rules on the bike-leg, which means that you have to push the pedals without the aerodynamic profit of riding in a highly organised, perfectly stream-lined (but because of the increased risk of crash quite dangerous) pack.

During a local race there usually are about 150 competitors. At the Powerman Zofingen, the yearly long distance world championship in Switzerland there may be up to 500 du-athletes, however.[2] Of course there is an environmental impact, but this seems rather low profile. Unlike during elite bicycle races, there are no preceding sponsor caravans during a duathlon. There are also no polluting cars with team managers and spare materials allowed during the race. Riders have to fix their own flats. Other than in elite cycle races, du-athletes are supposed not to litter. Racers have to leave their waste (banana peels, empty drink bottles, packaging) in designated areas. There is a time-penalty in case of violation. So generally the tarmac and the roadside do not suffer from the race. During the off-road mountain-bike version of the duathlon in the muddy autumn and winter-season there is some ecological pressure: worn tracks and probably of few broken twigs, but this will probably not be excessive.

As an individual competitor you might lower the environmental impact even more by travelling to the races by bicycle, train or by sharing a car. There may be a reasonable chance that you are a vegetarian, or only eat organic meat once or twice a week. But you have to admit; you do have more bikes—maybe up to six or seven—than you really need. And of course you do own more than just one pair of running shoes, say four. This overt materialism lessens your ecosophical joy, when you are honest with yourself.

---

[1] You, of course, know that the classic triathlon is the *conditio sine qua non* for a successful life as an endurance athlete, but since swimming demands an investment in technique and fixed training hours (especially when you haven't mastered the skill in your younger days), you keep postponing this switchover to what is considered to be the real thing.

[2] I will return to this specific (auto-)agonistic event in Chap. 7 *On Agon and Ecosophical Endurance: Finding your own Pace* by zooming in on specific casuistry.

So far for the shallower ecosophical considerations on the run-bike-run, now for something deeper. You will probably run and bike on different tracks (tarmac, gravel, dirt-roads, off the beaten tracks) under different conditions (sun, rain, wind, snow, uphill, downhill) at different speed and length (from a short recovery run to a 300 k bike race), during training sessions and race events. So you are doing quite well on Loland's ecosophical joy variety-scale, originally intended for sprint distances, but very suitable for outdoor endurance sports. You have to admit, however, you are probably rather limited in your bio-motor-abilities as a long distance rider and runner.[3] Frankly, you are more of a mono-maniac that simply pushes on. Thus you are not celebrating life in full motor variety. This is not very ecosophical, indeed. (But then again, what is the problem for the environment as such?)

Now, finally, for the overwhelmingly deep experiential part. What about your aspiration to acquire a full-hearted ecosophical mindset as a du-athlete, which is of paramount importance in Loland's proposed ecosophy of sport? Do you sympathise with the hypothesis that "[s]ymbiosis maximizes Self-realization potentials under conditions of limited resources"?

> What we deal with here is a process in which the apparently contradictory principles of unity and diversity constitute a synthesis or a gestalt in which they interdependent and mutually supporting. The key ideal of Ecosophy T, then, is this: 'Unity and diversity of life' (1996, p. 75).

Deep and symbiotic thoughts may pop up during an intensive lunch-break-run of an hour or so, or while absolving a long bike ride during the weekend. On your own or in a small silent group you occasionally even may sense (naturing) 'nature' and the interdependent diversity of being, unfiltered and at its purest. When perfectly pedalling or smoothly running without watching your actual speed or heart rate frequency, you finally might even come to understand Martin Heidegger's idea of *Gelassenheit*. This opaque and ominous concept is to be translated as 'releasement',[4] but it also incorporates characteristics such as calmness, complacency, resignation or withdrawal.

But even when competing with opponents in a race, there still may be left some idea of sympathy with the richly branched tree of life somewhere deep inside the competing athletes. In Loland's terminology this deep ecological sensation of experiencing unity in diversity is phrased in the following way: "The immediate joy of the 'interconnection between all there is' in competitive moments might be a microcosmos reflecting a macrocosmos in terms of an athlete's fundamental total view of the world"(1996, p. 85). This probably is what the Dutch cyclist Peter Winnen, twice winner of the Alpe d'Huez, a famous and forbidding climb *hors catégorie* in the Tour de France, meant when he wrote in a column that though during his winning ascents in 1981 and 1983 he wasn't able to think coherently anymore, let alone

[3] Although you recently might be in to all kinds of increasingly popular flexibility enhancing training programs such as yoga, tai chi, core-stability, Pilates, etc. This might ecosophically compensate somewhat for your one-sidedness.

[4] For a detailed account of how *Gelassenheit* got lost in translation—instead of deep 'releasement' it became shallow hedonist 'relaxation'—in the context of a discourse on ski-jumping, Cfr. Kreft (2010).

to carefully watch, enjoy and getting overwhelmed by the beautiful surrounding massif of the Grandes Rousses. Nevertheless, later he realised that the heinous Alpe had nestled itself in his body in an unconscious, subliminal way.[5]

This paradoxical experience of feeling scattered, lost and literally speechless but meanwhile also perceiving oneself as an integral part of the greater scheme of being similes what the American sport philosopher David Kilpatrick (2010) has called "muscle memory". The conscious mind cannot tell what the body knows—but still one knows. The Belgian philosopher Marc Van den Bossche reasons in a similar vein. Since our body is our primary access to the world, thinking, inexorably, begins with physicality. As a dedicated practitioner, for him endurance sport represents the art of living life fully.[6] Especially cycling has special benefits to offer when it comes to this. Pedalling against the wind or up the hill is a philosophical practice that connects him with the concrete world he tries to put into meaningful words. Therefore Van den Bossche proposes the following as his epitaph: "He thought by *vélo*" (2005, back cover, my translation).

Wrapping Up. What are the implications of the normative focus on ecosophical joy in my specific sport practice? In endurance sport there is a grey zone between utter exhaustion and a touch of the sublime that enables maximal Self-realization under limited but nevertheless stretchable conditions. The better you are trained, the potentially higher the ecosophical stakes. Whatever your result in a race, every time you push your limits there is sustainable growth.

3.2 Width and Depth

Second Ecosophical question
How should I, in my sport context, relate to norms for developing skills in width and depth, for playing to win and for applying only ecosophically sound sport technology?

'Depth' is a norm that you certainly hold dear as an endurance *aficionado*. You are always pushing the limits to get the best out of your running and cycling. 'Width' is more problematic, since, as argued before, your bio-motor skills are rather restricted. At the most you could argue that you practice two sporting disciplines. And, again,

[5] In the next Chapter, *Metabletics of Spinal Sport: When Poion meets Poson*, I will argue that the inherent repetitiveness of endurance sport even provides unique opportunities for enriching experiences. Once automated, running and cycling (and probably also swimming) enable unique possibilities of a high energetic unification with the surroundings. Once running or pedalling is ingrained landscapes will merge with the moving subject. "I point to an understanding of the kinaesthetic and sensuous experiences of the hybrid subject-object (in this instance 'the cyclist') as fundamental in rethinking how people live, feel, and ultimately create meaningful spatial relations" (Spinney 2006, p. 709).

[6] Other than Wilhelm Schmidt and Michel Onfray, who propose a rather hedonist philosophy of life at a much lower pace, Van den Bossche is a proponent of the eudaimonistic strenuous mood.

you probably practice your compulsory disorder also under different conditions, surfaces, surroundings and at different speeds, from sprint intervals, fartleks, extensive long distance runs and rides and racing just some 15% under maximum heart rate. So, there is some width in your skill development, if not of the exhilarating and agile kind that Loland seems to envisage.

'Playing to win' can be quadruple in your case.

1. Trying to end as high as possible in the total ranking.
2. Trying to end as high as possible in your age group.
3. Trying to be faster than at a previous event in a specific race.
4. Just trying to finish a race.

The last option probably applies to the manifold of people in a long distance event, such as running marathons,[7] cycling races,[8] or long distance duathlons or triathlons. This fourfold notion of winning and record-setting is wider, and thus more ecosophical than the uncompromising hunt down of opponents during a 100 metre dash.

In *Technology in sport: Three ideal-typical views and their implications* (2002) Loland proposes his 'thick' theory. He argues that sport performances do not just require equality of opportunity, but also should be "an arena for moral values and for human self-development and flourishing" (2002, p. 1) As long as there is no unnecessary risk for harm technology that requires effort en skills from athletes is permitted, say more sophisticated training programmes, lighter bicycles and running shoes.

The relation to technology as an endurance athlete is problematic. What is the optimum between gear and training? Imagine possessing a rather expensive time trial bike, not as expensive as some of your opponents, but more expensive than some rivals who are better than you. How to deal with this dilemma? Riding a high-end aerodynamic carbon bicycle increases the joy you experience during training and racing and might outweigh on a long-term basis the ecological costs of production.[9] But how to weigh the devilish dilemma of just enough and perhaps a little bit too much technology? History reveals that *Homo sapiens* inevitably uses ever more sophisticated tools. Humans meanwhile even have become technological themselves.

[7] In 1981 for the first time more than 400 Dutch ran a marathon in less than 3 h. In 1989 this number rose to 1593. In 2017 there were 15 times as many Dutch marathon runners as in 1981 (27,000). However, the number of sub 3.00 finishers (459) stayed about the same as in 1981 (Verkuil 2015). The same probably goes for du- and triathlons: more participants, lower average finishing times.

[8] Particularly so-called *gran fondo's*, cycling races with much altimeters, such as the Marmotte in the French Alps (5,000 altimeters), the Ötztaler Radmarathon in Austria (5,500 altimeters), the Marmotte de Pyrénées (5,600 altimeters) or the Swiss Cycling Alpenbrevet (6,800 altimeters).

[9] I will use the bicycle as a running gag from now, since it is the perfect grindstone for philosophical thinking on the (ecosophically) proper use of technology in sport. I will rely on the 'thick theory' Sigmund Loland suggests in *Technology in Sport: Three Ideal-Typical Views and Their Implications* (2001). For Loland, the thick interpretation: "does not just require equality of opportunity; its basic premise is that sport should be an arena for moral values and for human self-development and flourishing" (p. 1).

The Dutch philosopher of technology Pieter Lemmens even argues that man actually ís *techné*.

> [T]he human and technology are so inextricably and intimately intertwined that it can be argued that the human condition as such *is*—that is to say: is nothing but or boils down to—the technological condition. Man is the effect, the manifestation, of the becoming-technical of life, some four million years ago. Only with the rise of the technosciences since the Industrial Revolution, however, and in particular with the arrival in our times of the information and biotechnologies, does the true significance and real impact of man's technological condition become apparent (2008, p. 521).

One of Lemmens's main sources of inspiration is the work of Peter Sloterdijk, who argues that an assessment of the pervasive techno-sciences of the present has to build on a broad historical and cultural understanding of the ways in which science and technology have played a crucial role in the coming-into-existence of human beings as entities that are, to a significant extent, self-made and open to change. As already argued, Sloterdijk is a proponent of homeo-technology, a human- and nature-friendly usage of *techné*. We should change our lives for the ecosophical better. This also goes for sport. In *You must Change your Life: On Anthropotechnics* (2009/2013), Sloterdijk reasons that the call for *metanoia* has always resounded in human history. The preferential form, in which this radical reform of lifestyle or personal mitigation should be moulded, differs widely however, depending on in the era in which it manifests itself. In medieval times salvation was expected to come from monastery life as a spiritual exercise: a life in appropriate modesty, uniquely dedicated to God, scripture, abstinence and cultivating gardens.

Sloterdijk argues that as of the introduction of the Modern Olympics in 1896 the athletic ideal—already predominant in the philosophical athletism of the ancient Greeks—has become very influential once again. What we seem to have forgotten, however, is that it takes a lot of *askēsis* (training, practice, zealous repetition) to become a true athlete. The good life simply does not come without effort and pain. Those who explicitly, systematically and dedicatedly train themselves to become better in specific practices are the ones that really embody true existence. As already mentioned, also Loland acknowledges the (ecosophical) benefits of a well-performed training regimen, but only as a mean to a final end: "periods of monotonous hard work are sometimes necessary to reach deeper insight and values" (1996, p. 80). Sloterdijk is far more susceptible to the simple and self-referential nature of asceticism as such, however. "Practice is defined here as any operation that provides or improves the actor's qualification for the next performance of the same operation, whether it is declared as a practice or not" (2009/2013, p. 4).

Asceticism, then, is not just as a means to an end, say fitness, health, appearance or momentary lapse of exhilarating joy as a result of tough training sessions. In order to become resilient and up to our ecosophical tasks, we should learn to re-assess the quasi-Sysiphean nature of training as such. "The highest theorem of explicit training theories, then, is that ability subjected to persistent furthering tension produces, almost 'of its own accord', heightened ability" (p. 321). All this hardship will finally make us more resilient and up to the inconvenient ecological challenges we are facing.

For Sloterdijk—who used to be a runner in his younger days, but due physical discomfort changed to challenging himself on a bicycle—road cycling is one of the modern variants of ancient heroism. He believes that riding a high end carbon bike uphill brings us back to basics:

> Anyone can fight on flat stretches, but those who remain capable of fighting a duel on the worst of mountains already deserve to be called Hector or Achilles … Cycling represents for me a return to the primal man of the savannahs, who during a hunt spends the entire day running, and is constantly high in the process (Gorris and Kurbjuweit 2008).

Peter Sloterdijk, a philosophical practitioner. (Source: VPRO)

Cycling can help us to overcome the dichotomy of human beings and tools in a magnanimous and an up-to-date way. This is what the Danish philosopher Steen Nepper Larsen brings to the fore in his essay *Becoming a Cyclist* (2010).

> My consciousness is embedded in things and cognition is incarnated in a restless body. My *being-in-the world* is transformed to a *body-on-a-cycle-in-motion*, being able to do more than it knows. My identity is in a process of becoming an 'inter-being' between the bike, the experience and an ocean of interpretations. To cycle is an extended, mind stimulating rendezvous with and in nature (p. 29).

Nepper Larsen cautions that the environmental benefits of a cycling life cannot be justified by sheer utilitarianism in the long run. Rather, he justifies cycling philosophically, pointing out that in contrast to a modern sports utility vehicle, primitive bike technology "fosters an ecstatic-present-attentive being. One might say that I become bigger than my own flesh" (p. 30).

As a 6 foot 7 inch/2.04 metres tall person Nepper Larsen is condemned to ride a custom made steel bike. In (professional) cycling usually small and lightweight riders are the better climbers. Sloterdijk (quite tall himself) plays around with the term 'vertically challenged' by referring to the idea that we all want to 'go up', in one way or another (jumping higher, running faster, getting stronger and/or faster). In this specific case Larsen, paradoxically, would have done better to become smaller in order to climb the ever winking mountains easier. The very fact that Larsen's unsuitable physique still enables him to climb mountains of the *hors catégorie,* albeit at a non-elite level, renders him a Sloterdijk-inspired ascetic *pur sang.* He designs a vertical challenge of his own, so to say.

The thread of Sloterdijk's *You must Change your Life* is that we all can become our own personal Hercules or Achilles, regardless our specific physical set-up.[10] It takes a lot of training hours to finally give it a shot and head for the high lands. Nepper Larsen agrees with Sloterdijk, who reasons that mountains are not to be messed with. Either you climb them or you leave them alone. "Emanating from the mountains of the world, a vertical imperative hits the horizontal human. Pull yourself together, leave the lowlands and conquer the summit!" (Nepper Larsen 2010, p. 30).

What about the ecosophical soundness of your ascetological equipment? First you need a range of up to date tools to satisfy your vertical needs. In case of the duathlon this, next to proper clothing, a training bicycle, a racing bike, preferably a time trial machine, and perhaps a mountain-bike for the winter season. There is a wide price-range when it comes to bicycles. Depending on income, fanaticism and susceptibility for marketing you choose a specimen, say somewhere between 1,000 and 10,000 euro's. The more you pay, the less ecosophical you are, of course. But there is more to that. In order to attain ecosophical joy and maximized Self-realisation! you also need to take good care of your equipment, decidedly your bicycle(s). Since there is no mechanical assistance during a race, you need to be able to fix your bicycle yourself, at least at an elementary level. Thus you quite probably will oscillate between being a 'hip' or 'romantic' and a 'square' or 'classicist' rider, referring to Robert M. Pirsig's famous novel *Zen and the Art of Motorcycle Maintenance*: *An Inquiry into Values* (1974).

As a hip romantic (John Sutherland, the supporting role) biker of the un-motorized kind, during an extensive training session you will be overwhelmed by the stunning beauty of, say, the Alps. You then often even be flowing thoughtlessly,

[10] Cfr., for instance, *Only Cripples will Survive*, a chapter which deals with Carl Hermann Unthan, an armless violin player who performed all over Europe in the nineteenth century. "Unthan unquestionably deserves a place in the pantheon of reluctant virtuosos of existence" (p. 41). The nice thing about endurance sport is that motor ability is necessary but not decisive. Also relative cripples can blossom during a long distance endurance race.

not even aware that you were pushing the pedals quite hard. And sometimes you do take the unknown and unpaved road, which adds up to your ecosophical account, in terms of experiencing nature unfiltered and pure.

During tempo intervals and racing time, however, you can be as square and 'classic' as Phaedrus, the leading character in *Zen*, inspired on Plato's famous dialogue, and obviously Pirsig's alter ego. Then you are meticulous when it comes to bike care, and then you are the kind of cyclist that wants to know in advance where the road is going and how long and steep the ascents are. During a long distance run-bike-run, for instance the yearly World Championship in Zofingen, you may even be totally possessed by square and un-hip thoughts. Your bike must be in optimal form. You will force yourself to eat five power-bars, four bananas and three power gels during the 150-kilometre ride, during which you watch out for potholes and try to stay in heart rate zone 3, just under your turning point, as long as possible. And during the second, extremely hilly 30-kilometre run your motto probably will be: don't think of all the killing kilometres still to come, just accept the pain, keep going, stay hydrated, concentrate on your pace, lift your knees, etc.

Applying the 'only-ecosophically-sound-sport-technology-principle' to the duathlon eventually means: defining the right mean. In our case, the duathlon, the quest for the proper balance is about training harder (and thus increasing the risk of injury) or training more efficiently (which is especially beneficial for older athletes). One might also ponder over the question whether spending 1,000 extra Euros on a full carbon tri-spoke wheel outweighs the hypothetic gain of 2 min on 150 km. Also during the race the right mean plays a decisive role: finding your very own right pace, instead of chasing your opponents.

At a glance, such a high impact endurance race, which will probably take more than eight hours of continuous medium to high impact sporting activity, seems to be about *andreia*, the cardinal virtue of courage. Only the brave dare to try. But on second thoughts *sōphrosunē*, temperance, and *phronēsis*, practical wisdom, are the preferential virtues by far. Only knowing how to be confident but not foolhardy and an acquired routine of taking in food and fluid at the right moment will help competitors to survive in the long run. This sense for a healthy dose of restraint brings us back to the virtue ethical balancing act. Sport philosophers Kretchmar and Hwang (2010) argue that this striving for appropriate moderateness in sport—which at first sight seems to be more about exuberance, vehemence and limitlessness—is a reasonable restrictor, rather than a refuge of cowardice.

> [T]he golden mean is neither an arithmetic midpoint nor a lukewarm state of being noncommittal, tentative, or otherwise uncertain … It incorporates elements of each extreme but, with its location at the mean, avoids the evils that are found in the unbridled end points. Moreover, this mean is a state that takes into account the unique person and his or her specific environment (p. 103).

In conclusion: the width of the skills is limited, but there is a sheer endless reservoir of ascetological depth in the sports of long breath. With regard to the application of only ecosophically sound sport technology I would suggest the following rule of thumb: determine your own right mean between investing in high technology and training hours. For non-elite athletes buying a reasonable bicycle is a must, but it is probably better to invest more in training, quantitatively as well qualitatively, than investing a lot of money for a little loss of mechanic weight.

3.3 Close to Nature

Third ecosophical question
What can be done to promote sport training and competition in closeness to nature?

The third and final question is relatively easy to answer. Also serious contemporary endurance sport practitioners increasingly visit the gym: to run on a treadmill or to do a 'killer workout' spinning-bike in case of bad weather, or to strengthen muscles and improve core-stability, the latest hit in fitness-world. It is also true that triathletes often acquit their swimming sessions in indoor pools. But most of the running and cycling is still practiced in the open, thus in closeness to nature. Albeit that the 'natural' surroundings in which all this takes place may differ, ranging from city parks and the geometrical lanes of suburbia to the crisp countryside and desolate and tricky single tracks in a rugged mountain-landscape, as a dedicated endurance athlete you probably are committed to air as fresh as you can get.

When it comes to promoting sport in proximity to nature, endurance sport has another special benefit to offer: the deep length of becoming submerged in an inclusive lifestyle. While 'regular' sports such as volleyball, soccer, gymnastics or track athletics at a quasi-serious level already require a serious training investment, endurance sports outbid these by far. Just to be able to finish in a long distance duathlon or triathlon requires an investment of at least 20 h a week in the preceding 4 months. This usually implies a life-style in which sporting close to nature is not occasional, but integrated in daily life.

Endurance sport is process-minded rather than result-oriented, and thus implies a maximal potency for Self-realization! The holistic conception of a life fully lived in physical practicing, demands a positive stance towards a nature-oriented ascetology. Winning a race or improving your personal best is wonderful of course, but it is not a sufficient reason for submerging yourself in total asceticism. You do not have to be a masochist per se, but you have to like and to be able to see the deeper meaning of the hardship you have to undergo. Nature, then, should be considered as a warm bath for venting your asceticism, not as something oppositional, an ominous sphere that you have to conquer.

Triathlete Ken Glah, a 28 consecutive Ironman World Championships finisher in Kona, Hawaii, advises athletes to stay engaged by finding ways to enjoy training. Glah, who prefers training in his Pennsylvanian backyard, doesn't understand people who compete at the classic triathlon distance (3.8-kilometre swimming, 180-kilometre cycling, completed with running a marathon) and declare that they love racing, but hate training.

> I have no idea why they're in the sport. Even if you were to race 10 or 15 times a year, you're spending 95 per cent of your time training. If you don't enjoy the training, why would you do something you don't enjoy 95 per cent of the time, just to enjoy it 5 per cent of the time? (McDonald 2013, pp. 66–67).

To give the simplest of all possible answers to Loland's question of how to promote sport in closeness to nature: tell people of the joy of outdoor sporting, especially of the long and lonely rides and runs. Then perhaps refer to the Dutch psychiatrist and philosopher Jan-Hendrik van den Berg's[11] provocative and poetic plea for sturdy and single-minded solipsism. He argues for an ascetic life off the beaten tracks, "that demands full attention at every single step" (1973, p. 170, my translation). The metaphysical point that Van den Berg attempts to score is clear: do not yield to 'spinalism' or herd instinct, become your deepest self. Also sport philosopher Leslie Howe gives good ecosophical-ascetological reasons for sporting close to remote nature. She reasons that sporting in pristine surroundings has a greater capacity for deepening our self-knowledge.

> All sport has the potential to develop self-understanding and personal growth by offering various kinds of tests, but sport carried out in the remote and wilder places of the earth elevates this benefit because it commonly demands a higher than usual awareness of and response to risk, as well as presenting participants with a practical revelation of their relative significance (or lack thereof) in the natural environment. It can teach us not only about ourselves as human individuals, but also as humans placed in a wider world than the purely human. My claim is not therefore that remote sport is better than conventional sport as such, but that is does have some special benefits to offer (Howe 2008, p. 1).

While off the road events such as the *XTERRA: Global Off-road Triathlon and Trail Running Series* are increasingly becoming popular, many long distance endurance events still take place on paved roads. However, also tarmac has special benefits to offer when it comes to ecosophical nature-experiences. Magnanimous cyclist Steen Nepper Larsen contends that such confrontations add up to enlarging your often too cerebrally oriented self. "We have to listen to our vivid memory and create our own narratives. The work of the biking man's legs and muscles are his embodied thought" (2010, p. 38). Only experiencing and enduring the unpredictable curves of nature in a high physical state makes life worthwhile.

Loland suggests that the blossoming of as many talents as possible (width!) maximizes the ecosophical impact of sport. As already touched upon, this would make 'agility' sports such as skiing and all ball-games more prone to the radical green change Sloterdijk is striving for: "I am to develop into a fakir of coexistence with everyone and everything, and reduce my footprint in the environment to the trail of a feather" (2009/2013, p. 449). This personal awaking of responsibility for the greater good will ideally lead to a collective renaissance that is characterized by a "horizon of universal co-operative asceticisms" and by revaluing "the good habits of shared survival in daily exercises"(p. 452).

---

[11]Van den Berg's historical phenomenology will be put to the test in the following Chap. 4 *Metabletics of Spinal Sport: When Poion meets Poson.*

The one-dimensional cultivating of endurance (depth!) in stalwart daily exercises has special benefits to offer when attempting to enter the Elysian fields where deep mindsets result in perhaps 'shallow' but concrete arrangements. Transposing Loland's tentative ecosophical outline to Sloterdijk's tangible vertically challenged ascetological imperative for real life-change results a contemporary dark-green monastic rule. Aim for *metanoia*, train for sustainability, reach for your personal good. Understand that it takes a hell of an effort to get around. Run and (since this is a form of asceticism that does not require a lot of congenital motor talent, thus is within in reach of nearly everyone,[12] and is thus a perfect way of broadening ones scope in a sustainable manner for the crowd) foremost: cycle for life.

3.4 Metanoetic Olympism

After gradually drifting away to massive endurance, it is time to return to the initial frame of reference: exemplary elite sport, especially Olympianism. Commissioned by Olympic Fire, a task force of the Netherlands Olympic Committee-Netherlands Sport Federation, in 2011 I worked with a group of 24 students of the Top Sport Programme of the HAN University of Applied Sciences (Nijmegen, The Netherlands). It was our task to ponder over the ecological legacy of the Amsterdam bid for the 2028 Olympics. The rationale of this bid, which also included economical, social and, of course, sportive legacies, was that 100 years after the 1928 Amsterdam Olympics, the Amsterdam 2028 Olympics should become as sustainable as possible.

First of all we looked into the plasticity of the Modern Olympics. Are they really open for deep green change, beyond the usual light green window-dressing? A brief survey of the relevant literature learned that the Games are quite dynamic. Even more, many aspects of the modern Olympic Games we consider 'classic' were actually introduced during the 1928 Amsterdam Olympics. The Olympic Flame was lit there for the first time. (Although for security reasons the honour went to an employee of the national gas company, and not to an athlete, as became common usage during the subsequent Games.) The Olympic parade, with the Greek delegation at the start and the host country at the end, was also organised for the first time in Amsterdam (Arnoldussen 2008).

The Amsterdam Olympics were also the first with an official commercial sponsor (Coca Cola). And they also hosted the first sports icon, namely swimmer Johnny Weismüller, who soon after the games became a well-known movie-star, Tarzan, the archetype of the feral child raised in the African jungle by the *Mangani* a

[12] There is a growing market for so-called special need bikes. These suit people with all kind of physical impairments. Also at elite level there is more attention for sportspeople with a handicap in recent years. There are Paralympics and endurance sports competitions usually are also open for para-athletes. To put this praiseworthy development into a Sloterdijkian wording: almost everyone can relieve his or her sportive vertical tension.

fictional species great apes. Since then, athletics are contested on a 400-metre track and the games last for 16 days, instead of several months. For the first time there were women's athletics. German Lida Radke won the inaugural title in the 800 m in a world record time of 2 min and 16.8 s, which would last until 1944 (Arnoldussen 2008).[13]

Amsterdam Olympics 1928: An employee of the Dutch gas company lights the flame; "Me Tarzan, you Jane". (Source: Wikipedia)

Also with regard to Loland's plea for applying only ecosophically sound sport technology the 1928 Amsterdam Olympics provide an interesting perspective. A discussion arose when Englishman Jack London, who won the silver medal at the 100 m sprint, was the only athlete using starting blocks. The rest of the flabbergasted line-up still dug little holes in the gravel with their hands. Scientific testing proved that the blocks "were worth three or four hundreths of a second or about a third of a metre at the finish, not to mention the reassurance of knowing you wouldn't slip" (Harris 2009). This simple but effective novelty initially caused a thorough discussion on the use of technology and fairness. Because the rule book did not provide a possibility to forbid this technological innovation, the jury had to let London go. Nowadays, sprinting is unimaginable without blocks. These simple tools make athletes faster, and therefore add up to the total amount of ecosophical joy.

The most important conclusion of this (incomplete) historical retrospective is that the modern Olympics are flexible and open for change. Because of their adaptive nature and internationalism, the Olympics provide unique opportunities to open up for the final ecosophical end of the deep identification with all creatures great and small characteristic of an extended ecological Self. In the scenario for a potential

[13] Until then the 800 m for woman were abolished, because the women looked totally exhausted after finishing in 1928, which was qualified as un-esthetical. The exhaustion, however, was due to the fact that the participants were mostly sprinters, so not used to running 800 metres (Gratitude to Prof. Marjet Derks, Radboud University for this addition).

Amsterdam 2028 Olympics,[14] the students linked such themes as climate change, water management and energy production in various ways with solutions for stadiums, infrastructure and accommodation for athletes. There was also consensus on the idea that in the wake of the celebration of elite sport, the Amsterdam 2028 Olympic Games should pay special attention to recreational sport, social justice and public health. This resulted in the following ecosophical recommendations to Olympic Fire:

1. *Stadiums with different Post-Olympic Games use*. Stadiums should be constructed in a more sustainable and multi-functional manner: from temple for elite athletes during the Olympic Games to a home for the elderly, a school and a community centre afterwards. Even more, the students also suggested to revitalize the Olympic stadium in Amsterdam, which was built in 1928.
2. *Praise the all-round runner.* The students sympathised with Loland's suggestion for a reformed sprint scheme. Therefore they suggested the introduction of an ordinal ranking in a sprinting competition over several distances to let a larger variety of talents blossom, and abandoning the in the end untenable quest for unlimited growth in limited systems.
3. *Earth, water, wind, and fire: Olympics of the four pre-Socratic elements*. The students plead for using as many renewable resources as possible for (public and athletic) transport during the event. In the Dutch case means this would imply a special focus on the aqueous and the aerial, since we have plenty waterways and a good deal of fair wind. Whilst fire (carbon emission) and earth (land-use) should be used as sparsely as possible.
4. *Diving clean*. The students suggested to exchange the usual indoor pool for an outdoor Olympic diving contest in Rotterdam Harbour, where the water quality has improved exceptionally over the last years. Of course the weather conditions—in particular the wind-speed—should be closely monitored, but this unexpected *changement de décor* to outdoor air would have great symbolic value, they argued.

In addition to their attempts to socialize and 'ecologize' the Olympic Games, the students also studied and discussed Sloterdijk's ideas of asceticism and *Vertikalspannung*: our innate longing for verticality that overcomes the drudgery and monotony of everyday life. To determine the right mean between horizontal and vertical dimensions, we also assessed Sloterdijk's plea for revaluing the asceticism as a fundamental trait of human life. As already touched upon: Sloterdijk stipulates that roughly since the introduction of the modern Olympics in 1896 the athletic ideal has become influential again: "Whoever goes in search of humans will find acrobats" (2009/2013, p. 13). This provocative acknowledgement of a widespread sportive vein of course is a perfect starting point for reforming the greatest sport event on earth. To achieve truly metanoetical Olympics in 2028, the students made three supplementary ascetic-metanoetic suggestions to Olympic Fire:

[14] The Dutch government withdrew the bid to push back budget deficits.

1. *Introduce the Classic Triathlon.* At the Sydney Olympic Games in 2000, the triathlon was introduced: 1500-metre swimming, followed by 40-kilometre road cycling (drafting permitted, to make it more spectacular for the audience) and finally a 10-kilometre road run. This manifestation of *askēsis*-light should be replaced by the classic triathlon, as the ultimate expression of asceticism, which consists of 3.8 k swimming, 180 k of cycling with drafting and finally running a marathon. For the audience such a race, which lasts up to 8 or 9 h, would be less spectacular than the exciting Olympic distance, which takes less than 2 h. On the other hand, it should be noted that many people also love to watch long cycling races. The longer the race the longer the narrative potential, one could argue; long distance races may tell several memorable stories about the good life.
2. *Re-introducing the Pentathlon of the Muses.* To show that the overarching ideology of faster, higher, stronger also has an esthetical dimension. The group of students proposed to re-introduce the *Pentathlon of the Muses*. This refers to the competitions for music, literature, painting, sculpture and architecture, introduced at the 1912 Stockholm Olympics.

Also Pierre de Coubertin, the founding father of the Modern Olympics, entered the literature contest. Under the pseudonym Georges Hohrod & M. Eschbach he won the gold medal with his poem *Ode to Sport*, which from a philosophical point of view is a perfect specimen of applied virtue ethics. The nine stanzas respectively refer to the essence of life ("like the radiant messenger of a past age, when mankind still smiled"), beauty, justice, audacity, honour, joy, fecundity (rather proto-fascistically set to music as "destroying unhealthy seed and correcting the flaws which threaten its essential purity"), progress and, finally, peace.

> O Sport, you are Peace! You promote happy relations between peoples, bringing them together in their shared devotion to a strength which is controlled, organized and self-disciplined. From you, the young worldwide learn self-respect, and thus the diversity of national qualities becomes the source of a generous and friendly rivalry (Hohrod and Eschbach 1912).

Due to the growing difficulty of proving the amateur status of participants at the 1948 London games, the Olympic arts competitions were ended. After IOC President Avery Brundage, a firm opponent of professionalism, retired in 1972, the Olympic amateurism rules were steadily toned down to technicalities and lip service, until they were completely abandoned in the 1990s. This made the students conclude that there are good reasons to re-introduce the pentathlon of the muses.

3. *Running Free.* To open up new horizons for a sport that conquers technopia in an open and creative way we suggested to introduce *parkour* as a so-called demonstration sport. *Parkour* is the art of free-running, a sport where practitioners aim to move from one place to another, negotiating the obstacles in between, often in an urban setting. Still there is a strong connection to naturalness and maximal Self-realization! in this 'sport', which originates from the grim *banlieus* of Paris. In *parkour* there is no competition in the sense of who passes the finish line first. There is an element of aesthetics however, since 'competitors' discuss each

other's runs in terms of beauty, and exchange tricks. Another argument in favour of *parkour* is that no special facilities have to be built, which makes it even more viable for true ecososophy: rich ends by simple means!

> By dissolving the boundaries of the game's field of play, parkour has immersed itself into the environment like no other sport can. There are no longer stadiums, tracks, or determined pathways that force nature to concede to human power and privilege. Parkour integrates the environment, whether urban or rural, into the sport, and, in essence, allows nature to decide the rules of play. This fresh connection between athlete and environment creates many new perspectives of the world and the athlete's place in it (Greening 2011).

At a glance the suggestions the students made may look like a highly idealistic proposal for drastically reformed Olympics. On closer inspection, this historical and philosophical informed view is highly ecosophically and ascetologically justifiable. We must change our lives, also, or even particularly, in sport.

Running free for rich meaning. (Source: Deviantart)

3.5 Looking Back and Ahead: Cycling for Life

I started this chapter with an overview of recent attempts to develop an environmentally friendly philosophy of sport. Then I brought forward Loland's ideas on the ecosophy of sport. Although competitive sport obviously has negative side-effects—from aggression on and around the pitch to cheating and environmental damage—this doesn't have to mean that *homo ludens* has to give up his or her congenital urge to perform sporting activities per se. On a larger scale, which includes sports as a pastime or a health-enhancing technique for the masses, rather than the more or less organized and more or less serious competition, the intuitive appeal of striving for sport in closeness to nature becomes more complicated even under Loland's ecosophical outline. How close to nature can we get in sport? Abstract (deep) visions of nature sportspeople hold dear are often dissonant with their (shallow) actual behaviour.

Renouncing Loland's focus on elite sport and for developing sporting skills in width and depth, I thereupon concentrated on repetitive and rather numb mass-sports like running and cycling, and confronted these with Sloterdijk's provocative ideas on *askēsis* as the fit means for a radical change of our unsettled lives. I ended with some rudimentary ideas on how to make the Olympics more sustainable. Because of their broadness qua sports and internationalism, the Olympics are the best imaginable test-case for developing an able-bodied blue-print for a sustainable concept of sport.

Still, in order to reach for a more inclusive concept of a sustainable sport philosophy that is also relevant for the non-elite and not per se talented crowd, I turned to a more in-depth analysis to the kind sports that are suited for *homo repetitivus*: endurance sports, with a specific focus on cycling. This massive turn to the physically madding crowd is necessary, not just in terms of sport as a health-enhancer, but also in the sense of endurance sport as an apt tool for durable societal transformation. Sloterdijk is right: we must change our lives. It cannot be denied, the only fact of universal ethical importance is the ubiquitously growing insight that we cannot continue like this. We must decrease our deep ecological footprint, ideally to the weight of a feather. Ecosophically practiced endurance sport is a perfect device for arousing the *metanoia* we need.

Sloterdijk's intrusive call to change our life is daring and not very common in academic spheres. Philosophy usually is about more academic, abstract and 'high-brow' topics. I pick up the gauntlet, nevertheless, by reasoning that the noble art of the love for wisdom, to some extent, also can be directional. Practical philosophy has even special benefits to offer when it comes to proposing feasible alternatives for existing practices which result in environmental pollution, deforestation and the pursuing of instant satisfaction over a caring empathy beyond the inner human circle. Where economic reasoning and everyday anthropocentrism become less convincing, a practical and practicable approach to philosophy can help to overcome the stalemate between alarming scientific findings (urgently calling for a preferential change of direction towards a healthy planet earth) and a common, inclusive future for more species than just *homo destructor*.

Creating our personal upwardly oriented challenge, and leading what William James has coined life in "the strenuous mood"[15] is what makes being human worthwhile. We must ascend the mountains of life at our own pace. 2500 Years after Protagoras the ascetic bicyclical human should become the measure of all things. This is not a noncommittal metaphor, but a really existing alternative for the ubiquitous habit of motorised transportation. There is a fitting bicycle and a proper race or ride for everyone. Hopping on a bike certainly will help to become Sloterdijk's fakir of coexistence and to reduce our "footprint in the environment to the trail of a feather" (2009/2013, p. 449).

Thus, to bring the motto of *The Rider* to memory once again: "Non-racers. The emptiness of those lives shocks me" (Krabbé 1978/2016, p. 1).

[15] "[T]he deepest difference, practically, in the moral life of man is the difference between the easy going and the strenuous mood. When in the easy-going mood the shrinking from present ill is our ruling consideration. The strenuous mood, on the contrary, makes us quite indifferent to present ill, if only the greater ideal be attained" (James 1977, p. 627).

References

Arnoldussen, P. 2008. *Amsterdam 1928: de Spelen in Nederland*. Amsterdam: Thomas Rap.

Gorris, L., and D. Kurbjuweit. 2008. Philosopher Peter Sloterdijk on the Tour de France: 'The riders are just regular employees.' *Der Spiegel Online International*. http://www.spiegel.de/international/europe/philosopher-peter-sloterdijk-on-the-tour-de-france-the-riders-are-just-regular-employees-a-565111.html. Accessed 9 September 2017.

Greening, J.C. 2011. *A fresh perspective: Parkour and the infinite potentiality of the environment*. Paper presented at the annual meeting of the International Association for the Philosophy of Sport. 8–11 September, New York, Rochester.

Harris, T. 2009. *Sport: Almost everything you ever wanted to know*. London: Penguin Random House.

Hochstetler, D., and P. Hopsicker. 2012. The heights of humanity: Endurance sport and the strenuous mood. *Journal of the Philosophy of Sport* 39 (1): 117–136.

Hohrod, G., and M. Eschbach. 1912. *Ode to sport. V Olympiad*. Stockholm. http://library.la84.org/OlympicInformationCenter/OlympicReview/2000/OREXXVI32/OREXXVI32x.pdf. Accesses 12 November 2017.

Howe, L. 2008. Remote sport: Risk and self-knowledge in wilder spaces. *Journal of the Philosophy of Sport* 35: 1–16.

James, W. 1977. *The writings of William James* (ed. McDermott, J.). Chicago: University of Chicago Press.

Kilpatrick, D. 2010. *Poetics and the beautiful game*. Paper presented at the annual meeting of the International Association for the Philosophy of Sport, 15–19 September, Italy, Rome.

Krabbé, T. 1978/2016. *The rider*. London: Bloomsbury.

Kreft, L. 2010. Lost in translation: Heidegger and ski jumping in Slovenia. *Physical Culture & Sport Studies & Research* 49: 13–20.

Kretchmar, S.R., and J. Hwang. 2010. Aristotle's golden mean: Its implications for the doping debate. *Journal of the Philosophy of Sport* 37 (1): 102–121.

Lemmens, P. 2008. *Gedreven door techniek. De menselijke conditie en de biotechnologische revolutie (Driven by technology. The human condition and the biotechnology revolution)*. Oisterwijk: Box Press.

Loland, S. 1996. Outline of an ecosophy of sport. *Journal of the Philosophy of Sport* 23: 70–90.

———. 2002. *Technology in sport: Three ideal-typical views and their implications*. http://idrottsforum.org/articles/loland/loland_2.html. Accessed 9 September 2017.

McDonald, K. 2013. Long course for life – Ken Glah's 10 tips for Ironman longevity. *Triathlete Europe* (February), pp. 62–68.

Nepper Larsen, S. 2010. Becoming a cyclist. Phenomenological reflections on cycling. In *Cycling philosophy for everyone: A philosophical tour de force*, ed. J. Ilundáin-Agurruza and M. Austin, 27–39. Chichester: Wiley Blackwell.

Pirsig, R.M. 1974. *Zen and the art of motorcycle maintenance: An inquiry into values*. New York: William Morrow.

Reid, H.L. 2002. *The philosophical athlete*. Durham: Carolina Academic Press.

———. 2009. Sport, philosophy, and the quest for knowledge. *Journal of the Philosophy of Sport* 36 (1): 40–49.

Sloterdijk, P. 2009/2013. *You must change your life. On anthropotechnics*. Cambridge: Malden.

Spinney, J. 2006. A place of sense: A kinaesthetic ethnography of cyclists on Mont Ventoux. *Environment and Planning D: Society and Space* 24: 709–732.

Van den Berg, J.H. 1973. *De reflex: Metabletische tegelijk maatschappijkritische studie*. Nijkerk: Calllenbach.

Van den Bossche, M. 2005. *De passie van Marc Van den Bossche: Wielrennen*. Rotterdam: Lemniscaat.

Verkuil, D. 2015. *We lopen steeds massaler, maar ook steeds langzamer.* Nederlandse Omroep Stichting. https://nos.nl/artikel/2029707-we-lopen-steeds-massaler-maar-ook-steeds-langzamer.html. Accessed 2 November 2017.

Welters, R. 2011. Van Eyserbosweg naar Eyserbosweg: Fietsfilosofische plaatservaringen. In *Filosofische beschouwingen over plaats: Verbondenheid met natuur en landschap*, ed. R. van den Born, M. Drenthen, P. Lemmens, and T. van Slobbe, 51–70. Zeist: KNNV.

Chapter 4
Metabletics of Spinal Sport: When Poion Meets Poson

A healthy sport ethos will take note of the quantative, but will not let squelch out the qualitative value of the sport experience

(Heather Reid 2017, p. 164).

Abstract The previous chapter, which dealt with sport and the environment, implicitly ended with a quasi-Sisyphean take on endurance sport. He or she who takes up the challenge will eventually be able to control the mountains of life. According to the subsequent ascetic imperative we must immerse ourselves in diligent practice, create our personal upwardly oriented challenge and cycle for life. It is living in the strenuous and auto-competitive, but meanwhile also ecologically respectful mood that makes life worthwhile.

However, this emphasis on dedicated and necessary, but highly repetitive training-practices over the idea of sport as a playful voluntary attempt to overcome unnecessary obstacles (Suits, The grasshopper: games, life and utopia. Broadview Press, Ontario, 1978/2005), bears the risk of reducing sport to mechanic, soulless, un-reflective and un-critical activity. In this pejorative sense, then, sportive physicality becomes nothing but a matter of what Plato referred to as *poson*: calculative, quantitative measurability. Whilst the *poionistic,* qualitatively oriented quest for finding one's own subjective measure, ideally resulting in a harmonious and holistic sense of well-being is neglected.

To clarify and overcome the tension between quantity and quality, between dull calculable reps and rich, fully flourishing and meaningful life, in this chapter I will bring a specific brand of phenomenology to the fore: Jan-Hendrik van den Berg's *metabletica*. This doctrine of change, (*Metaballein* means 'to change' in ancient Greek) or 'historical phenomenology' is a rather daring attempt to unveil causality between at a glance unrelated events in a specific period. I will explain and apply

R. Welters, *Towards a Sustainable Philosophy of Endurance Sport*, Library of Ethics and Applied Philosophy 37,
https://doi.org/10.1007/978-3-030-05294-2_4

Van den Berg's disputed, but evocative method by paying a metabletical visit to two remarkable years, namely 1974 and 2010, which I will respectively assess as years of *poion* (quality) and *poson* (quantity). In terms of sport, 1974 appears to be a year of quality, of fully being-in-the-world. On the other hand, metabletically speaking, 2010 turns out to be a year of fixation on quantity, a one-dimensional 'calculative' understanding of the perfection of the self. At the end of the concerning paragraph I will critically assess this supposed watershed between the good and the bad take on sport.

Finally I will attempt to overcome Van den Berg's all in all unfruitful dichotomy of the 'reflexive spineless mass' versus the 'reflective critical individual' by arguing that his musings on automated, reflexive movement behaviour on closer inspection even have special benefits to offer when it comes to an ecosophical-ascetological understanding of sport. Only when every single step does *not* demand full attention anymore, one can look around and enjoy the magnificent scenery of life. Only after diligently putting in ascetic effort, one can attain the Elysian fields of ecosophical joy. Experiencing overwhelming *poion* always presupposes a robust amount of *poson*.

Keywords Metabletics · Quality · Quantity · Reflective spinalism

(The meta-metabletic attempt proposed in this chapter is actually a mid-term turn. In order to create a sustainable and robust philosophy of the enduring kind, in the next two chapters hermeneutics and pragmatism will be put to the ecosophical-ascetological litmus test. Thereafter, in the final two chapters, all sport-philosophical perspectives that have been mobilised so far will be brought together into a vivid comprehensive endurance pandemonium. This final practical and practicable synthesis is primarily focused on cycling, a reflexive-reflective which will appear to be the ultimate tool for building full ecosophic-ascetic embodied character for all. What the German philosopher Eduard Bertz was hoping for at the beginning of the twentieth century is still valid at the brink of the ecological doom in the early years of the twenty-first century.[1]

> The bicycle has been entrusted to 20th century man as a necessary completion to all other means for physical care, which in his estranged state from nature are not sufficient anymore. It has appeared not only to serve him as a helpful servant; it will also exercise its beneficial working as his physician (1900/2012, p. 54, my translation).

Other than Bertz, who was a proponent of cycling at a moderate and leisurely speed, I am an advocate of speeding up a bit, however. Eventually I will propose some idea of a golden mean between a narrow-minded high competitive speed and a slower pace that also enables pondering over the things of life in Chap. 8 *Epilogue: Turning in the Widening Gyre.*)

[1] Be it at in another state of aggregation and at a higher speed. Bertz thought that 12–15 kilometres per hour would be perfect speed for bike-riding. Because of technological innovations and vastly improved road-surfaces people meanwhile can ride at double speed quite easily. This higher speed also enables humans to travel longer distances in an environmentally responsible manner.

4.1 Controversial Ponderings

In the 1973 the Dutch psychiatrist and philosopher[2] Jan-Hendrik van den Berg (1914–2012) formulated a pungent critique of modern movement practices, such as travelling by train, aeroplane, or cycling, in *The Reflex: Metabletic Meanwhile Social Critical Study*. These are all fairly misbegotten mechanic activities, that have to be disapproved of because of their, literally, unconscious nature. Our contemporary movement practices run over our spine; they have become totally *posonistic*, as it were. Since collective, un-reflective herd-behaviour has taken the lead—especially in the way we travel, think of the reflexive dandling of people in a train, "they all waddle in perfect consonance"(p. 136)—*The Reflex* is an attempt to restore individualism. Go your own way! In Van den Berg's apodictic, Biblical style this goes as follows: "Broad is the path that leads to destruction, narrow the path that leads to life"[3] (p. 167).

Jan Hendrik van Den Berg: firmly against the tide. (Source: RD)

Before immersing ourselves deeper in his metabletic analysis of increasingly automated movement, a closer look at Van den Berg's fairly unique philosophical stance is necessary. First it should be noted that Van den Berg's metabletic method is strongly influenced by Edmund Husserl's phenomenology. Both authors argue that next to the objective world of the natural sciences there is a world of direct sensible subjectivity: a world of understanding, interpretation and meaning. They also agree on the phenomenological rule of thumb that the human perception of the world literally *ís* the world. Following this line of sensitive reasoning, Immanuel Kant was

[2] Although Van den Berg is not a philosopher in the full academic sense—he is a psychiatrist—his work has been strongly influenced by philosophers such as Husserl, Bachelard, Wahl, Merleau Ponty, and also Heidegger, whom he visited for 3 days in his *Hütte* at the Todtnauberg in 1947 (Zwart 2002, p. 19).

[3] My translation. All other citations of *De Reflex* are my translations as well.

wrong when he argued that it is impossible to know the *Dinge an sich*: things as such literally come to life by the act of seeing. *Esse est percipi*, to be is the perceived, following the well-known dictum of the subjective or empirical idealism of George Berkeley, that already has been mentioned with regard to the dynamic, 'naturing' nature of nature in the second chapter.

Husserl's method of eidetic phenomenological reduction starts from the conviction that one needs to see past the particularity of experiences. He is in search of iconic universality that lies on the other side of the concreteness of lived meaning. The idea of unveiling the phenomenological essence, or *eidos,* does not refer to some immutable universal or generalization about human nature of human life. For Husserl this would be yielding to the fallacy of essentialism. However, his quest for an underlying generic essence of the manifold of perceptual stimuli that we are exposed to, often leads to infinite discussions on the possibility of knowing what in everyday speech is called 'the world' at all. According to the Dutch technology philosopher Laurens Landeweerd (2016) Husserl's attempt to overcome the *noumena*, the thinkable dimension of the objective world, independent of the senses, by direct sensory access to the phenomena through eidetic reduction, results in notoriously hermetic, hesitant and long-winded writings. In his striving for objectifying subjective perception and for a return to the things themselves, Husserl has forgotten to materialise his ever fluctuating thoughts, paradoxically.

Also Van den Berg's metabletics can be seen as an audacious effort to reveal the 'hidden life of things' (Landeweerd 2016), I argue. While Husserl, however, drowns in inwardly oriented abstractions, Van den Berg addresses the reader directly with lively pictures. While Husserl looks at the glasses, Van den Berg looks through them.[4] His writings are anecdotic, tangible and concrete (Zwart 2002, p. 92); he overcomes the pitfall of unnecessary abstractions by using of imaginative and contrary casuistry. According to Van den Berg, there is a specific philosophical way of overcoming the dichotomy of being and appearing, of allegedly objective truth and the subjective eye of the beholder, of things and thoughts, namely: 'metabletics' or historical phenomenology. This turn to concrete historical events makes Van den Berg writings highly imaginative and tangible. When ancient man saw the world as the work of the Gods, there really were Gods, from Pan to Poseidon. And there was

[4] I borrow this witticism from the Dutch technology philosopher Peter Paul Verbeek, who used this metaphor during an introduction of Don Ihde, the godfather of the philosophy technology, in Nijmegen11 January 2018. Verbeek also argued in favour of 'material hermeneutics', a term which largely covers the approach of sustainable endurance sport I advocate in this study. According to Verbeek it is simply impossible to think of technology without a subjective human measure. Take for instance a telescope: "There is hermeneutics already in the instrument. It colours the way we see." Ihde as well as Verbeek are adherents of post-phenomenology. By means of their focus on the technological meaning and working of things—rather than Husserl's eidetic reduction (intuitively trying to identify the basic shape (*eidos*) of a perceived thing)—they better succeed in getting *Zu den Sachen selbst*, to the things themselves, I contend. In this hermeneutical-pragmatical vein I will approach the subject-matter of this study further on: the endurance sportsperson as a thinking technified thing in perpetual motion.

no such thing as 'the unconscious' before Sigmund Freud opened up this storehouse of unconscious desires, needs, and psychic actions in the 1890s.

The baseline of Van den Berg's metabletics is the idea that interpreting at a glance unrelated events in a certain period on closer inspection may reveal, or literally un-cover, striking similarities. Whereas traditional history looks at chronological relations and historical events as logical consequences of former events, metabletics is in search of at first sight un-related synchronous cultural and scientific changes. By paying closer attention to these synchronic phenomena, the true meaning and nature of historical or current developments can be brought to light. This often counter-intuitive, but simultaneously persuasive historical phenomenology discerns specific manifestations of respective 'Zeitgeists', to introduce a term that largely covers Van den Berg's metabletic investigation of specific eras. Zeitgeist is the link between, say, the rise of anatomy as a medical discipline and the emergence of a specific literary or art genre. For instance, what is the relation between the publication of William Harvey's *The Circulation of the Blood and other Writings* in 1628 and the architectural style of the Baroque period flourishing in the same period (Van den Berg 1996, p. 9)?

Starting point for the metabletic 'method' is an 'intriguing' incident that somehow catches the eye of the beholder, such as a major historical event, a literary or scientific publication, or a seemingly trivial event in everyday life (Zwart 2002). From there Van den Berg looks for different, synchronous events. Crucial for Van den Berg's metabletic method is non-disturbance of the material on hand. There should be no pre-selection and no single fact, however seemingly unimportant, should be excluded in advance: "in principle everything is important" (Zwart 2002, p. 32, my translation). All facts that pass the metabletic litmus-test are justifiable. Synchronous facts of importance are those which, for one reason or another, somehow attract the attention of the metabletic beholder.

This selective perception goes for initiatives at the brink of a new type of scientific research, publications that are highly discussed or, on the contrary, have been wrongfully neglected. As well as events that will have important political implications, or even handicraft innovations, such as a high heel under a ladies shoe. "Each selection is arbitrary to some extent, but this is not a problem" (Zwart 2002, p. 169, my translation), because what is important is the link or pattern that emerges.[5] All things considered Van den Berg's method provides a unique opportunity to really the understand the human measure. It is the quest for similarity between the seemingly un-similar itself, the metabletic thought-experiment as such, that creates meaning in a technologically convenient but meanwhile erring and cold world.

While often praised for his daring and original style, Van den Berg frequently has been criticized for his radical conservatism and anti-egalitarianism. As of *De Reflex*[6]

[5]This is also acknowledged by Bertha Mook, a Canadian expert on Van den Berg: "Metabletics provide a uniquely interdisciplinary approach through the analysis of simultaneous events to identify patterns in human experience. Most central to the metabletic method is that, while the world of science is constant, the landscape of human existence is continually changing and causing humans to change" (2009, p. 26).

[6]To fully understand Van den Berg's central point of criticism the in every-day speech often

(1973) metabletics, which had been strictly historical so far, will also serve as a means to scrutinize and criticize actuality (Zwart 2002, p. 23–24). From now on the reactionary nature of Van den Berg's thoughts becomes explicit. Quintessential is the subtitle: "a metabletic study that at the same time is critical of the social structure". All things considered this concise study is a malevolent critique of contemporary egalitarianism and its associated thoughtless and uncritical herd behaviour.

Van den Berg contends that only after the discovery of the knee-jerk, or patellar reflex, by the German neurologist Carl Westphal in 1871, human life could become totally automatic. This successfully switching off the testee as a conscious person resulted in B.F. Skinner's behaviourism, which, according to Van den Berg, is a political and social program that wants to prove that all men are equal, rather than scientific psychology. That is to say: behaviourism wants to prove that humans are modifiable through a simple scheme of stimuli and responses (Van den Berg 1973, p. 161).

One of the running referential gags in *The Reflex* is the loathing of Jean Jacques Rousseau's *moi common*, the generalized, common I, which, as was argued earlier, can be linked to Heidegger's *Das Man*, the indeterminate and elusive 'They'. According to Van den Berg, Rousseau's perfidy 'trick' is that the individual has to exchange his or her singular I for a neutralized common I: "he has to become 'everyone'" (p. 113). Who travels to Italy, hopes to find real Italians instead of an indiscriminate *moi commun,* Van den Berg contends. Before the Industrial Revolution started in England in the middle of the eighteenth century, handicraft work determined specific national identities. Each country or region literally *manu*-factured specific tokens of craftsmanship, such as traditional costumes, wooden shoes or a specific type of scythe, a local sausage or a French carriage. After industrial machinery had been introduced, work gradually turned into soulless automated production. And so did mankind: eating tasteless Hamburgers and flying in non-descript aeroplanes, with their empty heads spinally nodding all in the same direction.

According to Van den Berg, the general I that Rousseau's puts forward in his *Du Contrat Social* (1762/2012) depraves humankind from its individuality. Rousseau imagines that humans endorse an unanimous contract, making their community a moral and political association. They thus create a social body that constitutes a public person (the *moi commun*). This comes with a price, however, Van den Berg argues. The sediment of the *volonté general* is a depersonalised ego, a spinal life and a general will. In the social contract the colourful and concrete person evaporates into grey and tasteless communality. This is what Van den Berg tries to counteract by means of his historical phenomenology.

> One should furthermore realise to what extent fast transport and fast communication, both enabled by the industry, already have seeded the *moi commun*. The whole world became, after McLuhan's expression, a *global village*, complete with the world-gossip of radio and television. With regard to this development nationalism has a protective meaning (p. 114).

In Van den Berg's case, this plea against Rousseau's *moi commun* leads to a straightforward conservatism: globalisation, equalisation and unification are in-humane,

---

neglected difference between 'reflexive' (trusting on ones reflexes, running over the spine) and 'reflective' (thoughtful, deliberative, contemplative, so the very opposite of reflexive) is crucial.

therefore we should return to pure and pastoral pre-automatic life. He seems to suggest that we should look for national and local identity, rather than diluting ourselves homeopathically in uncanny globalism. We should strive for undefiled and pure life instead of going with the spinal flow. In order to revitalise authenticity, Van den Berg advocates to switch consciousness on again, so that humans become real, vivid persons again, who are more than just a brainless bundle of automatic reflexes. In order to do so, the author literally wants to take us back to pre-reflexive times.

> So many things are running over our spine. Driving a car, cycling, hiking. Sporting achievements run over our spine. He who controls movements, performs automatically. Automatic movements control our existence. The last two centuries however, more automatic movements are performed then before. Before that time, people hardly racticed sport. There were no cars. No bicycles. There was no machine, that urged us to acting automatically. Automatism has become in control over our actions (preliminary note).[7]

All moving in the same spinal direction: "If , in other words, the car would largely disappear from public life, then in neurophysiology vitalism in all honour, and with an open mind, could recover itself" (Van den Berg 1973, p. 147)

Van den Berg particularly disapproves of those who practice sport in groups. Concerning cycling in a peloton he sighs: "The reflex discerns herb nor tree, knows

[7] From a grammatical and stylistic point of view Van den Berg's written Dutch is sometimes somewhat peculiar. He often writes so-called elliptical sentences; sentences which are not complete from a grammatical point of view. This style is often used by journalists in order to please and/or seduce the reader. Van den Berg's writing is furthermore also extremely subjective, teleological and moralistic. Carefully avoiding the 'I think style' he still addresses the reader directly with N1, authorative, straightforward seductive and suggestive statements, hardly ever underpinning his findings by strong scientific-statistical empirical evidence.

no nature; the reflex has no land, no homeland" (Ibid. 117). As already touched upon, Van den Berg wants to lead mankind off the beaten track again.

> The masses do not make decisions. The crowd is driving car. Leave this path, contemporary. It is still possible. Chose the path that is too narrow for a spinal mass. You should go alone, by two, by three. Only in a small group, or in solitude, one can find life. He who is in search of life, leaves the broad road, which leads to doom and destruction, and chooses the narrow path, that demands full attention at every single step (p. 170).

Although Van den Berg's provocative thoughts may be regarded as reactionary in the literal sense, a specific reading may also open up ecologically progressive horizons. His call to leave the beaten tracks and choose the narrow path (away from the well-paved avenue towards the egalitarian I), can also serve as a plea for a life that is less focused on material gain and more oriented on qualitative growth. Interpreted in this vein, Van den Berg's critique even concurs with the anti-hedonistic undercurrent and 'non-anthropocentrism' in Arne Naess's deep ecological philosophical skeleton, outlined in the second chapter. True ecosophy strives for rich ends, brought about by simple means; life is about original quality, not about industrially reproduced quantity. Life is not an industry but a craft.

Although departing from a different political *Weltanschauung*, a different guiding perspective on how to lead the good life, both Naess and Van den Berg underpin the need for a 'back to nature' lifestyle. This longing for undefiled naturalness shares a common ground with Rousseau's 'theory of 'the natural human' and the hypothetical 'state of nature' in which there is no need for competitiveness or overconsumption. However, whilst Naess above all is pre-occupied with the positive ecological effects of a befitting return to naturalness, Van den Berg hopes for a moral reorientation of disoriented humanity as a result of the return to pre-spinal, more authentic conditions.

This 'reactionary' disposition explains Van den Berg's aversion to Rousseau's concept of the *moi commun*.[8] For him this idea of a 'generalised I' leads to colourless inequality instead of a renewed appreciation of difference. According to Van den Berg, Rousseau has paved the way for inhuman communism. The *moi commun* operates in a reflexive rather than a reflective manner. The *patellar reflex* has become the paradigm of human behaviour, and is therefore the key target of Van den Berg's conservative pamphlet against 'progressive' spinalism.

Paradoxically, however, one might argue that his retrograde escapism implicitly pays tribute to Rousseau's back to nature principle. Leave the broad and levelled path, and choose the older, more primitive and bumpy option; the one that leads to a supposedly more bucolic, unspoilt life. The problem with supposed authenticity is

[8]"Each one of us puts his person and all his power in the supreme direction of the general will; this renders each member as an indivisible part of this general will. At that very moment, the particular person of each contractor by means of this act of association produces a moral and collective body composed of as many members as the assembly consists of, which receives from this same act its unity, it's me, its life and it's will" (Rousseau 2012, p. 12, My translation).

that it is always reconstructed in hindsight. Detailed historical research tells us that idealised past times usually were at least as grim as our present, or even worse. Van den Berg's idealisation of a pre-reflexive past resembles Rousseau's state of nature. An important difference is, however, that while Rousseau seems more aware of the fact that the natural state is hypothetical, Van den Berg takes us back to concrete and carefully selected historic events, such as the "end of an authentic inside", which, according to Van den Berg, occurred at the end of the Baroque period (1973, p. 151).

From the perspective of mainstream sport philosophical research it is important to realise that both impetuses—the leftist as well as the rightist—entail a call for a return to nature, towards a less hedonist human stance, and thus to more sustainability. This post-romantic pragmatic plea for a befitting contemporary ecological modesty, that pays tribute to the flourishing of as many biological species as possible, which is definitely in the air, recently also has been taken up by leftists critics of the detrimental effect on earth of *Homo sapiens*.

According to these critical thinkers, during the so-called 'anthropocene', nature is becoming drastically malleable by human technological intervention. This term, introduced by the Dutch atmospheric chemist and Nobel Prize winner (1995) Paul Crutzen,[9] has also been embraced by contemporary philosophers such as Peter Sloterdijk and the French philosopher Bernard Stiegler. For reasons of urgency the latter prefers to use the term "Neganthropocene", an era in which humans should become less omnipresent, less omnipotent and less destructive.[10] The Neganthropocene should become an age not of appropriation, but of appropriate human modesty. For example: enjoying nature nearby by means of a bicycle-ride rather than travelling to an eco-resort in Costa Rica by plane.

Following the main line of reasoning of Van den Berg's critical assessment of reflexive, 'cold' modernity, we now must ask ourselves: how to turn the anthropocentric tide towards a more positive, more ecosophically oriented direction? How to find new sporting arrangements in the Anthropocene, the era in which mankind has taken the unforgiving lead over life on earth? How to reconcile and integrate qualitative and quantitative (especially bicyclical) efforts to create a sustainable philosophy of sport? And: what is the feasibility of Van den Berg's metabletic critique of 'spinalism' for such an inclusive philosophy of sport?

---

[9] The Russian geologist Alexei Pavlov coined the term Anthropocene already in 1922. Together with the ecologist Eugene Stoermer, Crutzen re-introduced the term in its current meaning in 2000 (Close e.a. 2016, p. 33, footnote 18).

[10] "A new critique of anthropology, both philosophical and positive: did this not become necessary from the moment in 2004 when we saw Claude Lévi-Strauss on television admitting that he is preparing to depart a world he no longer loves? If anthropology cannot *account* for this becoming that so disheartens the anthropologist, does it not thereby lose its legitimacy, just as has occurred to those philosophies that pretend to be unaware of such questions? In other words, what becomes of anthropology in the Anthropocene era? My thesis is this: it becomes a neganthropology, and it must contribute to the advent of the Neganthropocene" (Stiegler 2014, p. 1).

4.2 Shifting World-Views: From *Poion* to *Poson*

Despite the circumstantial evidence in favour of metabletics provided above, the question that still remains is: why applying this 'method', so highly inductive, anti-positivistic and subjective per se, to matters of sport, especially those of the during and durable kind?

There obviously is a critical methodological weakness in metabletics, or historical phenomenology. Van den Berg begs his ominous questions. Things cannot be discovered out of thin air, thus there must be something in the air, he argues inductively. He more or less operates a permanent naturalistic fallacy: when there is an indeterminate uneasiness about the estranging and de-humanizing effect of modern technological culture, there apparently simply must be something wrong with human morality, since humanity, after all, enabled this becoming of *techné*. Being aware of its questionable shortcomings, but still speaking in defence of this 'method',[11] I argue that metabletics potentially can bridge the gap between quantitative, rule-oriented, objectifying, nominal and narrowing views on sport and subjective, un-filtered, broadening perceptions of the direct environment, ideally resulting in an increasing ecological-ecosophical awareness of sport.

As already stipulated, Van den Berg is reluctant when it comes to explicitly defining his method. Zwart concludes that he only refers to this by means of occasional hints and methodological asides. It seems as if reading the writings themselves provides the clue to understanding the basic principles of his historical phenomenology (Zwart 2002, p. 64). From a more humanities-oriented perspective it can be argued that metabletics, or historical phenomenology, overcomes the classic cumulative linear historiographical approach of representing history as intertangled chains of influences. The attentive observer reads the files in a different manner, italicizing what authors seem to mention as an aside, and turning it into a fact of metabletical importance (p. 65). By means of this imaginative technique, apparently isolated facts become an integral part of a coherent meaningful unity; they begin to form a lively *picture* (p. 69). This unconventional approach is necessary, Van den Berg argues, because we live in disquieting times of discontinuity, which makes it hardly possible for us to imagine the "appalling continuity and stability of the past.... A chronic uncertainty characterizes our acting", such as Zwart portrays Van den Berg's deep *Unbehagen* (p. 69). Since we are at the brink of doom, the inductive approach of metabletics, looking for linking patterns, is not only justifiable but also necessary, Van den Berg seems to argue.

Besides providing an imaginative narrative for forging seemingly randomly chosen facts into a comprehensive picture, which would make metabletics a hermeneutic

[11] 'Method' is etymologically tributary to the ancient Greek *hodos meta*, which means 'above the road'. Although not methodical in the current quantitative statistical sense that dominates the social sciences, Van den Berg's metabletic method is mixture of a helicopter view from above and a road-trip at the same time. With respect to *De Reflex* this means a detached view on society as regressing to reflexivity blended with very concrete and precise, but arbitrary random pictures.

frame of reference,[12] Van den Berg subsequently develops as decidedly normative position, as is especially visible in *De Reflex*. This transition of metabletics as a form of *Verstehen* to a form of societal criticism is necessary to clarify the negatively connoted disruptions he discerns in human history. The word 'bicycle' could only be coined as of 1870, when the roads were paved, modern industry emerged, and humans were both forced and enabled to move reflexively. And the widespread uneasiness in culture, voiced by Van den Berg by Zwart, could only come into existence because time was ripe in 1973, the year in which *De Reflex* appeared.

The Anthropocene, one could argue, is the final result of this shift towards reflexivity (instead of reflectivity), giving rise to the decidedly normative question 'how are we to live'? Can metabletics (as a geneaology of reflexive mobility) help us to unravel the relation between ecosophy and sport? Even more, since 'Anthropocenic science' (climate change research, biodiversity research, etc.) results in the imperative that a radical change of lifestyle is inevitable, metabletics may provide extra ammunition, because it combines a historical analysis of the origins of the current crisis with a therapeutic appeal to individuals. Sloterdijk's plea for human-friendly homeo-technology that mimics the dynamics of nature instead of domesticating it, is supplemented with a focus on lifestyle, a willingness to live a life far less material and much simpler, without becoming a Luddite.[13] "Van den Berg's oeuvre is not a curiosum, it is strikingly topical.... His works deserves rehabilitation" (Zwart 2002, back cover).

I will now test the sportive value of a metabletical approach by applying it to two remarkable, metabletic years, namely 1974 and 2010. Both can be regarded as years of transition, but both exemplify quite contrasting sportviews, or even worldviews. To further elucidate the contrast, I will make use of Marc Van den Bossche's analysis of 'good' and 'bad' professional cycling practices in terms of the Platonic distinction between *poion*, or quality, and *poson,* or quantity (2005, p. 26). In short, my metabletic analysis will show that whereas 1974 exemplifies a focus on *poion*, 2010 rather reflects a shift of focus towards *poson*. Van den Bossche elaborates his distinction with the help of concrete examples. For him, the Kazakh cyclist Alexandre Vinokourov ('Vino') is the prototype of a *poionistic* rider, always attacking, and

[12]Although there are interfaces between phenomenology, hermeneutics and metabletics, Mook argues in favour of attributing a specific weight to metabletics. "The metabletic method certainly involves a detailed reading of historical events and texts, including documents, novels, and paintings. Van den Berg did not, however, specifically address hermeneutics and did not aim for a systematic hermeneutic reading and interpretation of texts. Instead, he incorporated his historical insights and interpretations into his conceived metabletic project. Despite the fact that phenomenology, hermeneutics and metabletics share some common characteristics, they remain distinct qualitative research approaches in their own right. They can be seen as mutually enriching and complimentary approaches aimed at disclosing the rich yet mysterious meanings of our lives as lived and experienced." (2009, p. 28). I will take up the hermeneutic perspective in Chap. 5 *Ascetic Practices, Hermeneutical Cycles and Ecosophical Endurance* and broaden it from texts to more directly experienced phenomena, or practices, such as sporting experiences.

[13]A group of early nineteenth century English workmen named after Ned Ludd. Their main aim was destroying labour-saving machinery as a protest. Luddites argued that automation destroys jobs. In a broader sense a Luddite is one who is opposed to technological progress.

never calculating. This makes Vino a 'romantic' or 'hip' rider in the sense of Robert M. Pirsig's cult-novel *Zen and the Art of Motorcycle Maintenance* (1974),[14] already mentioned in the previous chapter. According to Van den Bossche, Lance Armstrong, on the contrary, is the *posonistic* rider par excellence. He is a decidedly calculating and utilitarian rider, who pays a lot of attention to bike mechanics. Thus, Armstrong resembles Pirsig's main character and alter ego Phaedrus,[15] who is 'classic' and 'square'.

Building on these perspectives, I will first turn my attention to the year 1974, a year which, notwithstanding the spinality of sport at large and cycling in particular, still stands out as a qualitative or *poionistic* year. In other words, the inevitable spinal dimension of sport/cycling did not yet erase but rather enhanced its experiential quality. This was different, I will argue, in the *posonistic* year 2010, a year in which the inherent spinality of sport/cycling indeed resulted in a fixation on the quantitative. How to reconcile these two observations? In other words, how to address the fact that, notwithstanding the historical phenomenological association between sport/cycling/endurance and spinality, the former is not inevitably linked with a posonistic obsession with quantity? As a philosopher of sport, but also as a philosophical practitioner, I will explain how 'spinal, mass reflexivity' and 'individual reflectivity' can go together by introducing the idea of 'reflexivity re-enabling reflectivity'.

4.2.1 1974: A Year of Poion

Why 1974? What was of metabletic importance of the year that was preceded by the publication of *De Reflex*, but also the year in which Arne Naess coined the term deep ecology and the year of the first triathlon ever (in its extended version the ultimate manifestation of asceticism, I argue) in San Diego?[16]

First of all it has to be pointed out that during this year sport was in the air, both in the public and in the philosophical realm. In 1974 an academic journal uniquely dedicated to one of the activities listed by Van den Berg as spinal activities was founded: the *Journal of the Philosophy of Sport*. But 1974 is also the year of the much-debated soccer world cup final between Germany and The Netherlands, as well as the year in which Eddy Merckx won his fifth and last Tour de France and became the first winner of a world championship organised outside the old world,

[14] Already touched upon with regard to the use of ecosophically sound technology in the previous chapter.

[15] A reference to Plato's dialogue *Phaedrus*, probably written somewhere between 370 and 360 B.C. *Pheadrus* is a polyphonic dialogue in which Socrates discusses the matters of life lying under a plane tree with Phaedrus. The dialogue deals with the art of rhetoric and how it should be practiced, but also with topics as *metempsychosis* (the Greek tradition of reincarnation) and erotic love. In his fictionalized autobiographical novel Pirsig refers to his past self (before undergoing electroconvulsive therapy after collapsing in academia) by using the name Phaedrus.

[16] This event started with running (6 miles), followed by a bike-leg (miles) and ended with swimming (500 yards), so the reverse order of the present triathlon (Johnstone 2017).

namely in Montréal, Canada. In addition, 1974 is the year of Robert M. Pirsig's bestseller and cult-classic *Zen & the Art of Motorcycle Maintenance*, and, finally, 1974 turns out to be the year of the last meeting of the *London Positivist Society*. Can a metabletical connection be discerned between these synchronous, but apparently disparate facts? I will argue that this is possible indeed.

In the first edition of the *Journal of the Philosophy of Sport*, anno 1974, Scott R. Kretchmar attempts to demonstrate the "close relationship between day-to-day experiences (including sense perceptions, feelings and understandings) and phenomenological analysis" (1974, p. 119). He does this against the backdrop of Husserl's *The Crisis of European Sciences and Transcendental Phenomenology* (1936/1970). During the very years when the natural sciences, particularly physics (e.g. quantum physics and Einstein's theory of relativity) were enormously successful, Husserl in this unfinished book addresses the historical adverse development between deep understanding and factual explanation: "The positivistic reduction of the idea of science to mere factual science" (p. ix). The 'crisis' of science refers to the loss of its meaning for life. Husserl finally attempts to overcome the (modern) opposition between "physicalistic objectivism" and "transcendental subjectivism" by drawing renewed attention to the "life-world as the forgotten meaningful ground of natural science" (p. 48).

> Necessarily, we believe in *the* world, whose things only appear to us differently but are the same. [Now] have we nothing more than the empty, necessary idea of things which exist objectively in themselves? Is there not in the appearances themselves a content we must ascribe to true nature? (p. 23–24).

In order to lever the tension between the idea of things and the things as such, Kretchmar brings forward his own sport experiences in a phenomenological manner. He interprets sport primarily as a matter of *con-testari*, of testing-together. Competitive sport, then, becomes a competitive tool for enriching people's lives. Sport adds spiritual value, rather than reducing people to nothing but spinal vehicles. There is more to sport than organised and rule-bound competitive encounters between contestants, moreover. Kretchmar also acknowledges the meaning of testing oneself. When climbing a mountain on your own you may succeed or fail in reaching the summit. "But in any case, I have not won, lost or tied anything (a characteristic of a contest). I have merely acted in relationship to a problem of hindrance and accomplished certain things" (1974, p. 130).

Kretchmar's personal but (at least for dedicated practitioners) recognizable ponderings prove that beyond sheer reflexes, sport potentially also may denote a deeper reflective meaning. The analysis of his personal sporting experiences results in a life-style which oscillates between Van den Berg's pensative individual, separated from the indiscriminate crowd, and the spinal, non-reflective sportsperson (the mobilised Heideggerian "They"). In between contest and test, Kretchmar therefore suggests a third way: "Perhaps I might invent a new activity called 'exer-sport' in which one alternately and in unique ways lives exercise by itself and contesting in the singular game"(p. 131). Metabletically speaking, it cannot be coincidental that in 1974 a pensative philosophy of sport settled down in an academic journal. There must have been something in the air.

If we move over to actual high-profile sports practices in 1974, we notice something similar, especially in the realm of cycling. Although the Tour the France of that year enlisted a herd of cyclists well-equipped by the bicycle industry and engaged in spinal movements, there seemed to be more at stake than Van den Berg's automatic, unreflective and, literally, headless (acephalic) cycling peloton,[17] discerning herb nor tree. In 1974, professional sports—from soccer to cycling—also serve as carriers of societal values and may eventually open up broader and unexpected horizons. The following two path-breaking examples can qualify as food for thought n this context. Both were the outcomes of a process that started earlier, around 1967 in fact, known as 'the summer of love', when thousands of youngsters flocked in the neighbourhood of Haight-Ashbury, San Francisco. In 1974 it became clear that the critical anti-consumerist and anti-war sentiment that flourished in the avant-garde hippie-community during the late 1960s infected the Western masses in a critical reflective manner and even became noticeable in the way in which (at least some) sportspeople perceived their vocation.[18]

First metabletic stop. Until the soccer world cup final against West Germany in 1974, the Dutch played "the beautiful game",[19] beating all their opponents so far by their superior technical, tactical and even aesthetical skills. According to David Winner's *Brilliant Orange – The Neurotic Genius of Dutch Football* (2000) the supreme march of the Dutch squad through the tournament until the final marked a new era in professional soccer. For Winner the rise of this superb and nonchalant game-playing also was tantamount to a new, less subservient, unvarnished attitude towards life, the kind of attitude that had already been expressed in Nietzsche's *The Gay Science*: because God is dead, new and broader horizons occur.[20]

Especially Johan Cruyff, the technical[21] and tactical genius who seemed omnipresent, giving tactical advice to his team-mates, and who was famous for his apodictic, often enigmatic comments on life in general and football in particular, exemplifies the critical spirit of the still the resonating hippie-era. Think for yourself, don't believe the generally accepted. "What is speed? Often the sports press mixes up speed with insight. Look, if I start to run a little earlier than somebody else, it just looks as if I am faster."[22] Cruyff refutes Van den Berg's critique of spinalism in sport

[17] Peloton in Dutch also refers to the, military connoted, 'platoon': organised mindless and will-less herd-behaviour *in extremis*.

[18] Because of his conservative nature Van den Berg probably would disapprove of this analogy, but, again, I argue that Van den Berg's critique of spinalism also opens up for a leftist reading.

[19] Cfr. Torres 2012.

[20] "Indeed, we philosophers and 'free spirits' at the news that the 'old God is dead', as illuminated by a new dawn, feel that our hearts here flows with gratitude, amazement, premonitions, expectation - finally appears us the horizon again free to set himself that he is not bright, at last our ships may run again, run out at all hazards, any risk of the knower is allowed again, the sea, our sea is back open, maybe there was still never as an 'open sea'" (Nietzsche 1954, Band 2, p. 206, my translation).

[21] 'E.g. the so-called 'Cruyff turn': "The turn was a way of tricking an opponent into thinking that a pass was to be played but then turning behind his defensive lunge and dribbling away"(Mumford 2014, p. 190).

[22] NCRV-gids (Dutch radio and tv-guide), June 12 1982, nr. 24, my translation).

beforehand: "I hate doing things that I do not do consciously."[23] Today Cruyff's often enigmatic and apodictic sayings, often with a mangled grammar, even are used in management books and leadership trainings.

Now to the story of the final match. In the second minute, before any German player had even touched the ball, following a solo run, star player and captain Johan Cruyff, often compared to a ballet dancer, was brought down by the sturdy Uli Hoeneß. Although the foul was committed just outside the penalty box, referee Jack Taylor granted the Dutch a penalty, scored by Johan Neeskens. Be it with a stroke of luck, since Neeskens kicked half into the ground, and so unintentionally misled goalie Sepp Maier. The Dutch became overconfident after this quick lead, not to say arrogant.[24] The Germans canalized their irritation and struggled to recover. They equalized by a penalty scored by Paul Breitner. Gerd 'the bomber' Müller scored the winning goal. By this unexpected win the *Mannschaft* once again affirmed the famous tongue in cheek quote by former professional football player and BBC commentator Gary Lineker: "Football is a simple game. Twenty-two men chase a ball for 90 min and at the end the Germans always win."

In the collective memory of the Dutch, the 1974 world cup final usually is considered as the ultimate failure to establish a new spirit in football: the old obedient working ethic won over the new frivolous game which revaluated all existing values. A more detached view on the game however learns that the German win was deserved, because the Dutch simply were not able to express their supposed superiority. Moreover, the German player Breitner probably represents the anti-authoritarian post summer of love spirit more than Cruyff, with his charming political naivety. Breitner was a rebel with a cause (Kok 2004, p. 328). At the age of 18, he abandoned university studies (pedagogics, sociology and psychology) in Munich to turn professional with Bayern. He excelled both as a marauding left back during the 1974 World Cup and later as a dominating midfielder who demonstrated exceptional tackling ability, a thunderbolt shot and natural leadership qualities.

Breitner also was famous for his unconventional approach to life off the field. He was regularly seen bringing Mao's little Red Book to training sessions and openly declared his sympathy for Che Guevara. He was also more honest than any other player with regards to doping in football. In the aftermath of the World Cup Breitner declared that during the tournament the laudatory stories about the Brilliant Orange irritated him, since the Dutch were behaving arrogant, especially Johan Cruyff, who was, above all, playing for himself.

Now for a second sportive stop in 1974: professional cycling. Eddy Merckx won his fifth, and last Tour de France and his third and last world championship, in Montreal, Canada, for the first time out of Europe, the traditional cycling continent. Because of his persistent habit to try to win every single race he started in—uncommon in professional cycling, which always had been a partially *posonistic* game of win and take, of negotiating and calculating—Merckx's nickname was 'the

[23] Tussen Barend en van Dorp (Dutch radioshow with a special interest in sport, March 1995).

[24] According to journalist and writer Auke Kok the Dutch were "loose, brutal and indifferent" (2004, p. 327, my translation).

cannibal'.[25] He was a *poionistic,* non-calculating rider *pur sang*, one who overcame the existing, 'treacherous' values for a quest for quality.

Though 1974 was a very successful year for Merckx, his sovereignty started to show cracks, however. The wear and tear was beginning to show. He still won *La Grande Boucle* 8 min and 4 s ahead of Raymond Poulidor. However, the Frenchman defeated the hitherto all-devouring cannibal twice in the Pyrenees: at St-Lary and on the Tourmalet. Moreover, his compatriot Michel Pollentier beat Merckx at Orléans, just before the last stage to Paris, by 10 s in the time trial, the hardest discipline, in which Merckx had been invincible so far. In 1974 Merckx gradually started to ponder over his indomitable sovereignty.

God on wheels

[25]The story goes that the often used word cannibal (literally: man-eater, figuratively: blood-thirsty, cruel person) stems from the Spanish 'canibal', a corrupted amalgamation of 'caribal', Caribbean Indians, who were accused of cannibalistic habits, and 'can', Galician for dog (*canis* in Latin), another species that consumes peers. Pastor Samuel Purchas, however, claims in his in 1625 published *Hakluytus Posthumus or Purchas his Pilgrims, contayning a History of the World in Sea Voyages and Lande Travels, by Englishmen and others* that cannibal in the Caribbean language stands for 'courageous man'. Both etymological explanations perfectly match the phenomenon Eddy Merckx.

During the next year's *Tour de France* he was punched by a spectator in the abdomen during the climb of the Puy de Dôme, and he therefore couldn't close the gap to the leaders. At stage 17 misfortune struck again. Merckx crashed heavily. Although injured, he quickly remounted his bicycle and continued the race with a bruised hip and knee and a broken jaw. Although struggling to breathe, he refused treatment. Despite his injuries he still somehow managed to continue to ride hard. By the end of the stage the cannibal finished third, just 2 s ahead of Bernard Thevenet, his French challenger, who eventually would win this Tour, 2 min and 47 s ahead of Merckx, who became second.

Eddy Merckx, the impervious and relentless rider, who never gave anything away and seemed to be a pre-programmed winning machine, became a notorious doubter. The classic invincible Greek God landed on earth and became a fence-sitter, who due to the before-mentioned severe crash constantly was fiddling with his saddle position and seat height.[26] In 1974 Merckx transformed from a reflexive machine to a reflective person, attentive at every single step/pedal stroke. Due to his decline as an unhesitating Hercules, Merckx paradoxically became fully human.

Another event that adds substantial qualitative metabletic weight to 1974 is the publishing of *Zen and the Art of Motorcycle Maintenance*: *An Inquiry into Values* by Robert M. Pirsig. This novel, already mentioned in the previous Chapter, nuances Van den Berg's insinuation that modern ways of moving forward are nothing but surrendering oneself thoughtlessly to the spinalism of (in this case motor-) cyclism by introducing the dichotomy of cycling 'hip' and 'romantic'[27] versus riding 'square' and 'classic'. John Sutherland, the supporting role in the novel, as a hip or romantic biker, doesn't care too much. He simply hopes for the best with his bike and when problems do occur he becomes frustrated and irritated and is forced to rely on professionals to repair it.

In contrast, Phaedrus, the leading character and Pirsig's alter ego in this highly autobiographic novel, has an older motorcycle that he is usually able to repair himself by using rational problem-solving skills. Phaedrus is an academic philosopher, who after a personal crisis was hospitalized in a mental institution, and after his

---

[26] "[T]he fact that I continued in the 1975 Tour de France after I crashed definitely did shorten it. My build-up to that race had already been problematical, and actually I wasn't in the best of health when I started it. But after the crash, in which I fractured my cheekbone, I suffered like you cannot imagine possible. I could not take in anything but liquids. I had to race on empty. I had to continue for the sake of the race, for honour and for my teammates. They depended on my prize money. Remember that I still finished second. What I should have done, looking back, was pay my riders what I would have earned out of my own pocket and left the race. Then maybe with my strength rebuilt I could have been competitive in 1976" (Cycling Greats 2010).

[27] 'Romanticism' is a problematic term. It usually implicitly refers to characteristics such as idealistic, poetically dreaming and somewhat detached from real life out there. From a historic perspective one has differentiate between early romanticism, the highly provocative style of the *Sturm und Drang* period, which lasted from 1760s to the early 1780s. Individual subjectivity and extremes of emotion were expression in reaction to the perceived constraints of rationalism imposed by the Enlightenment. This period was followed by high romanticism (approx. 1800–1850), a period of a ripened romantic artistic, literary, musical and intellectual movement and late romanticism (1850–1890), the rather domesticated variant of the Biedermeier-period, which tended to realism again.

discovery became a writer of user manuals. This new profession taught him to think in a different, less abstract manner. In order to succeed, a manual has to observe the human measure. Tightening a bolt exactly right, not too tight, but certainly not too loose, is also a matter of the right feeling, developed over time and through meticulous practice.

In his novel Pirsig on the one hand characterizes riders that are interested mostly in Jungian *Gestalts*, vague emotions and feelings, and on the other hand those who seek to know the details and master the mechanics. The resulting distinction between (poionistic) hip riders and (posonistic) square ones is indebted to Jung's psychology of types: classical introverts versus romantic extraverts. Eventually, however, Pirsig demonstrates that rationality and a Zen-like 'being in the moment' not only can harmoniously coexist during a truth-seeking road-trip, but also may propel each other towards a better, more flourishing life. By means of a subtle and integrative critique of 'mechanic life' he touches upon Arne Naess's rule of thumb of simple means that ideally result in rich ends. Phaedrus knows how to handle a motor-bike in such a way that it becomes a reflexive extension of his reflective self. In the novel, he finds life on a path that is perhaps not very narrow but still is quite deserted; a paved one that provides elbow room for *poion* ánd *poson*.

Another event that draws the attention of the metabletic beholder is the official end of philosophical positivism. 1974 is also the year of the last meeting of the *London Positivist Society*. This progressive club was founded in 1867 and modelled on the positivist society established by Auguste Comte in 1848 for the purpose of promoting what by positivists was called the religion of humanity. "The meeting place of the London society was decorated with the busts of the saints of humanity. Regular services were held where a positivist liturgy was observed, positivist hymns were sung and positivist sermons were delivered" (Long 2003, p. 74). In 1974 there apparently is no longer room for the idea that logical or mathematical deductions and empirical findings are the exclusive sources of all authoritative and scientific knowledge.[28]

There are some other events in 1974 that draw the attention of the metabletic beholder. For instance the invention of the 'Rubik's Cube' by the Hungarian sculptor and professor of architecture Erno Rubik. Each of the six faces of this 3-D cube is covered by nine stickers, each of one of six solid colours: white, red, blue, orange, green, and yellow. An internal pivot mechanism enables each face to turn independently, so that the colours can be mixed. For the puzzle to be solved, each face must be returned to have only one colour. Serious mathematics becomes a game for the masses. Furthermore, it cannot be without metabletic meaning that the Swedish pop group ABBA wins the Eurovision Song Festival with their hit Waterloo, which was the first step on their path to worldwide fame. The single became a number 1 hit in

[28]As opposed to Comte's 'shallow' positivism, Husserl considered his philosophy of pure experience as *truly* positivistic: "We take our start from what lies prior to all standpoints ... we are the genuine positivists" (cited in Drabinsky 1993, p. 226). The silent liaison between William James's radical empirism and hermeneutics and phenomenology will be put to the test in Chap. 6 *Continental Pragmatism: Enduring Life in the Strenuous Mood.*

many countries, and went on to sell nearly six million copies, making it one of the best-selling singles of all time. And finally also Ivan Illich's mentioned ode to the bicycle as the ultimate tool for elevating mankind in a sustainable manner anno 1974 cannot escape the hawk's eye of the historical phenomenologist.

Again, since in metabletics historic disruptions do not happen *ex nihilo*, these events contain hints concerning what was brewing in the air in 1974. It cannot be without meaning that this is the year of the last meeting of the *London Positivist Society*. The year of the foundation of the *Journal of the Philosophy of Sport*. The year of Cruyff (or at a closer look rather Breitner) introducing un-conventional and critical thinking in football. As well as the year of Merckx and Phaedrus annex Pirsig reconsidering their lives. At first sight there indeed is something mechanical and 'reflexive' in bicycles, the Rubik puzzle, the basic technique of soccer, professional cycling, the ABBA rhythms, in positivistic logic, etc. But at the same time 1974 shows that in the folds and margins of this mechanic, reflex-based world, quality and reflectiveness can still resurge. In other words, the common thread in these events is that they synthesize acting reflexively (*poion*) and acting reflectively (*poson*) into a comprehensive 'frame' of mind (to use a cycling term).

The similarity in these at a glance isolated events is that they synthesize acting reflexively and acting reflectively into a more comprehensive framework of understanding *(Verstehen!)* life rather than simply explaining *(Erklären!)* it schematically. Metabletically speaking, 1974 may be considered an *annus mirabilis*, a wonderfully 'thick' year in sporting terms. "Here, sport is interpreted as a social practice with its own characteristic norms, values, and internal goods that are again linked to more general human virtues" (Loland 2001, p. 5). Set against a posonistic, mechanical backdrop, 1974 turns out to be a year of *poion*, of human flourishing of sport as a post-positivistic, sensitive and rich means for fully being-in-the-world.

4.2.2 2010: A Year 2010 of Poson

Now, for a second metabletic stop, we will focus on the year 2010. To give away the conclusion beforehand: this *annus horribilis* will turn out to be a year of full-fledged *poson*, of a regression to the average mean. 2010 Is a year of spinalism and single-handed techno-optimism.

Positivism is in the air again. According to Comte's classic positivism, the physical world is supposed to operate according to absolute physical laws, and so does society. If this all-embracing idea of technological progress will be put properly into practice, this leads to the foundation of an un-metaphysical, science-based *Religion of Humanity,* Comte *cum suis* contend. Due to the enormous technological progress that science has enabled over the last decades—from biotechnology and materials science to information sciences—the following series of apparently disparate observations point to a re-instalment of positivism, a positivism 2.0 in 2010. This unshakable belief in the overwhelming beneficial effect of technology—from the solution of environmental issues to the relief of all human suffering—results in a *Religion of Trans-humanity*.[29]

First observation. In 2010, the *Technological Innovation in Sport Award*, sponsored by the *British Sport Industry Group* ("Where sport meets business") was conferred to the Mongoose cricket bat. According to the manufacturer, this new token of high-end sport innovation "reduces the length of the splice to increase the hitting area and give you the biggest, most sizzling sweet spot in cricket". This means that the noble art of cricket becomes easier to practice and thus accessible for a larger spinal audience. Hitting the ball straight is nothing but a matter of using the right tool. It is a material choice rather than an acquired, internalised practise.

Second observation. In 2010 the new sports centre of Radboud University Nijmegen (Netherlands)[30] was delivered. It was inaugurated as the *Gymnasion*, as a reference to the classic Greek schools for the development of body and soul. On further consideration, however, this high tech sports center appears to be anything but a Platonic *akademeia*, an Aristotelian *lykeion* or a Cynic *cynosarges*: schools of life where the development of motor ability and athletic skills went hand in hand with philosophical discussion and moral education.[31] In the twenty-first century the *gymnasion* has become a crystal palace, a narcissistic cocoon where people run on treadmills, exhaust themselves on spinning bikes and practice all sorts of fitness, equipped with heart rate monitors and output meters. Van den Berg would doubtlessly have considered it to be the ultimate manifestation of spinal mass in optima forma. It evokes an indiscriminate *moi commun*, which neither chooses the narrow path, nor discerns herb or tree.

It has to be admitted: also in 2010 there is still a phenomenon called 'teamsport'. In the early eighties, however, there were 60 soccer teams on 9,000 students in the internal competition at Radboud University. In 2010 there were just 15 teams on a population that rose to 16,000 students. And yes, it also has to be recognized that in 2010 there also was a tendency towards 'moving on music', sporting—or more adequately: 'exercising'—in a group, where beats convey all kinds of more or

[29]The Humanity+ Transhumanist movement states the following: "Humanity stands to be profoundly affected by science and technology in the future. We envision the possibility of broadening human potential by overcoming aging, cognitive shortcomings, involuntary suffering, and our confinement to planet Earth. Interesting the transhumanist movement has also an ecosophical perspective: "We advocate the well-being of all sentience, including humans, non-human animals, and any future artificial intellects, modified life forms, or other intelligences to which technological and scientific advance may give rise" (Humanity+ Transhumanist Declaration). I consider this as a classic technological fix which cunningly avoids the issue of human responsibility in the Anthropocene. Instead of acting an attitude of a post-Heideggerian *Gelassenheit* seems more appropriate. I will elaborate on this active ascetic passivity in Chap. 8 *Epilogue: Turning in the Widening Gyre*.

[30]My *alma mater*.

[31]Socrates to Glaucon: "What I should say therefore is that these two branches of education seem to have been given by some god to men to train these two parts of us—the one to train our philosophic part, the other our energy and initiative. They are not intended the one to train body, the other mind, except incidentally, but to ensure a proper harmony between energy and initiative on the one hand and reason on the other, by tuning each to the right pitch." (Plato 1981, p. 176).

less coordinated movement, such as steps, Zumba, Pilates, and so on. These health practices, however, bear more of the characteristics of a narcissistic way of working on the perfection of the body than enjoying the grace of each other's movement. Instead of discussing the topics of life afterwards in Socratic dialogues, participants have a quick shower and go their own way. All things considered it cannot be but concluded that in 2010 at the Gymnasion 2.0 of Radboud University, mass sport has become totally *posonistic*. It has become a sheer instrument for the measurable perfection of the self.

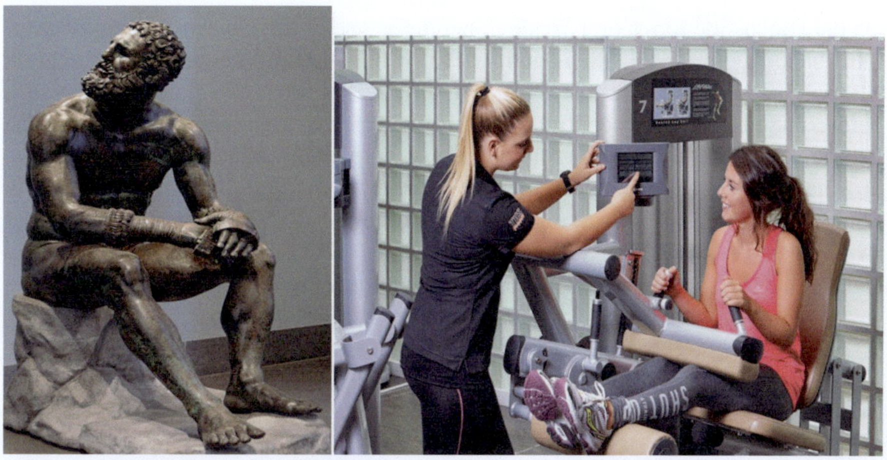

From deep thinking to shallow measuring

Also a metabletic look at subsequent elite sport-events seems to confirm the *poson* hypothesis. In 2010 the Netherlands reached, once more, the final of the World Cup in South-Africa. This time they lost to the Spaniards by 1–0. The loose, sublime and over-confident 1974-style was replaced by result-driven cynical pragmatism.

> The style of playing (with an extreme focus on results) has been characterized by many as 'non-Dutch'. According to The Independent: 'It is hardly the stuff of David Winner's "Brilliant Orange" we are describing here; nothing like the 'totaalvoetbal' perfected by Johan Cruyff under Rinus Michels's leadership in 1974 in which all the Dutch players were so completely gifted that they could interchange positions in the 4-3-3 formation which the side displayed to the world (Van Hilvoorde et al. 2010, p. 234, note).

Of course, after two lost finals,[32] it seems fair to argue that the Dutch had good pragmatic reasons to play their defensive and extremely result-oriented game. Sufficient reasons to shift from *totaalvoetbal* (total football) to *resultaatvoetbal* (result football). But, as Torres (2012) rightly argues, there is more to football than simply scoring or avoiding goals at all costs:

---

[32] In 1978, 4 years after the defeat by West Germany, they lost the final against hosting country Argentina by 3–1.

> It is not too venturesome to maintain that even those who believe that what reigns supreme in football, or any other sport for that matter, is that the pragmatism of winning and thus favor conservative or downright negative football, given the chance would prefer to win in a style that embodies the way of the beautiful game (2012, p. 309).
>
> To put it differently, *how* sport is conducted is important, both morally and aesthetically. For example, contestants should try their best in competition not only to provide a genuine challenge that accurately measures their relative ability but also to advance the best interpretation of their sport. This, in turn, honors the virtue of integrity defended by the interpretivist philosophers of sport (p. 313).[33]

On the Dutch side top-scorer Wesley Sneijder is one of the most prominent players. Skillful and aggressive, with a good shot in both legs, well-known for his exuberant tattoos, his marriage with a TV-star, but also for his often ostentatious annoyance before the camera when confronted with the negative style of playing during the tournament. But Sneijder is hardly the person with a view on life at large in the line of Johan Cruyff, or perhaps rather Paul Breitner, as suggested in the previous section. In Nietzschean terms Sneijder is more of a consolidator of the existing order (winning at all costs!) than a player that re-values all existing values. He is a touchy and highly paid servant who carries out pre-calculated tactical tasks rather than a sovereign game-changer who overcomes all existing, constraining values.

Also several events in professional cycling in 2010 are of metabletical importance. The Tour de France was initially won by Alberto Contador, a romantic, intuitive and aggressive rider, the last patron of standing pedalling,[34] nick-named *il pistolero*. However, Contador was later revealed to have failed a doping test. Because of this, the *Court of Arbitration for Sport* decided in 2012 that he lost his results from 2010, declaring Andy Schleck, who originally was 39 s behind, the new winner. Schleck, who later also turned out to be a doping sinner, is more of the calculative, premeditated and strategic kind, only attacking when the time is right.

Of further metabletic importance is that the Norwegian powerhouse Thor Hushovd won the men's elite road race during the World Championships in Melbourne, Australia in a devastating sprint. He simply was the one who was able to produce the most Watts, 1,600 to be precise. Of course sheer power always has been a necessary condition in professional bicycle sprinting. The fact, however, that physical power now unreservedly is translated into measurable units makes 2010 explicitly posonistic.

---

[33] The interpretivist or broad internalist take on sport will be put into perspective in the next chapter.

[34] "Standing pedalling allows you to apply more force to the pedals than is possible seated, because you can rest your entire weight on the driven pedal, and, even more, by pulling up on the handlebar, you can push the pedal with more than your actual weight... but is this a good thing? Pedalling that hard is very stressful to the joints and to the bicycle, and usually involves a level of effort that cannot be sustained aerobically (that is, you will get out of breath). Unless you have unusually good form, it also tends to involve a fair amount of thrashing from side to side, which is a waste of energy. The added stress flexes many parts of the bicycle, and the energy required to do this flexing is not usually recovered when the parts straighten back out." (Brown undated).

Also of metabletic importance is the fact that in 2010 it has become compulsory to wear a helmet in professional cycling and nearly all riders wear sunglasses. It is hardly possible to discern a grimace of a rider in distress. The pain that was all over Merckx's face during the mountain stages of the *Tour de France* in 1974 is anno 2010 concealed by plastic. The rider sees a filtered world and the audience sees a pain- and sun-filtered rider. From now on the truly human is hidden under a veil.

When things went wrong: taking refuge in high technology and hiding emotions. (This is not what it at first sight may look like. Behind the screen there is a real person. This is Pieter Aben, the Dutch national time trail champion at the masters (55+). He is an amateur in the true sense: he dedicates his life to cycling without getting paid.)

Also the use of earplugs transmitting tactical orders from the team-manager and so-called power-meters is now common practice. In 2010 the peloton has become an indiscriminate and remote-controlled machine, a shiny high-tech *perpetuum mobile* without lined faces, a pre-programmed *homunculus* "discerning herb nor tree" (Van den Berg 1973, p. 117), fixated on the display that show the wattages they produce on their steering wheels and waiting for commands from the team-manager in their earplugs.

While in 1974 tacit knowledge on preferred dietary habits was commonplace in the peloton, in 2010 positivistic science has taken the lead. Food intake, for instance, can be customised to various age-groups. In the *Journal of the International Society of Sports Nutrition*, Chen et al. in 2010 publish an article entitled *Arginine and Antioxidant Supplement on Performance in Elderly Male Cyclists: A Randomized Controlled Trial*. This study indicates that tailor-made supplements potentially improve exercise performance in endurance athletes beyond their prime.[35]

Due to the increased 'scientification', also at a non-professional level, endurance sports such as cycling are no longer about listening to your body and developing a certain feel for the right measure (*poion*), but simply about following the biochemical imperative (*poson*) provided by high-tech sport science. As a result of technological innovations and sophisticated scientific training programs, seriously performed endurance cycling in 2010 has become a carrier of instrumental values. Winning a race is not about being daring and relying on your intuition. Winning has become a matter of applying the proper scientific calculus. Kretchmar's tentative internal quest for meaning has turned into human malleability by state-of-the-art biophysics.

From a metabletic perspective it must be concluded that 2010 is the year of the massive break-through of technology and of banning the faith in one's body. The well-known Delphic maxim "know thyself" has changed into "re-design thyself". Sport has become totally quantifiable, nothing but an instrument for the measurable perfection of the self. 2010 Is an *annus horribilis*, a year of *poson*, of sheer computation. It is a year of ultimate *omphaloskepsis,* that is: navel-gazing. Whereas ancient *omphaloskepsis* was a means to an end, striving for a sense of unity with the Gods, its modern variant boils down to sheer self-oriented Narcissism.

4.2.3 Evaluating the Metabletic Watershed

The basic disquieting historical phenomenological intuition of Van den Berg's *De Reflex* is that humans increasingly act in automated ways, without interference of the critical and conscious mind, notably when it comes to transport and movement in the era of ever-increasing industrialisation and automation. According to him this tendency has to be stopped, because it diverts us from our original destination: being truly human, being pensative and going solo (or duo or trio, but that's the maximum, according to Van den Berg). Since the stakes are that high—during the time of writing he identified the detrimental effects of posonistic self-estrangement—2010 certainly passess Van den Berg's critical reflexive litmus test: *poson* has absorbed

[35] Zwart (2000) points at the ascetic undercurrent in medieval food ethics: "Whereas ancient Greek food ethics concentrated on the problem of temperance, and ancient Jewish ethics on the distinction between legitimate and illicit food products, early Christian morality simply refused to attach any moral significance to food intake. Yet, during the middle ages food became one of the principle objects of monastic programs for moral exercise (*askesis*)" (p. 113). Following this line of reasoning, as of 2010 sport *dietetics* become profane asceticism, one might argue.

poion. But the normative imperative is still there: choose the path that is too narrow for a spinal mass, and search for real and reflective life beyond the mechanic iterations of the madding crowd.

In contrast, although similar mechanistic and reflexive tendencies were already at work, 1974 stands out as a miracle year, not because of the quantitity of sportive highlights, but because of their meditative reflective poionistic quality. Apparently, the authentic poionistic individual could still overcome the posonistic tendencies. Observed through the lense of historical phenomenology, however, 2010 turned out to be a year of utmost mechanic reflexivity. Anno 2010 subtle and tentative *poion* has given way to brute and straightforward *poson*.

In the post-2010 era of increasing reflexivity, contemporary sport can hardly be considered an appropriate tool for carving out an ecosophical-ascetological stance. Sport does not seem to result any longer in an open-minded *Sitz im Leben,* a life setting, an attitude which lets many flowers blossom co-existentially: a comprehensive and peaceful attitude, which still enables to vent our inborn vertical tension by climbing the mountains of life, but at our own pace, as proposed at the end of the previous chapter.

But this should not be considered the inevitable final word, I argue. The lesson to be learned from 1974 is that, in the folds and margins of posonistic society, exceptional reflective experiences can still be attained. Notwithstanding Van den Berg's metabletic intuition that the macro-world is moving in spinal directions, an ecosophical-ascetological understanding is still something worthwhile to pursue, albeit against the grain. Van den Berg's descriptions of sport as spinal, reflexive and automated movements running over the spine, to some extent seems to have taken over critical self-reflection. What he seems to overlook, however, is that there is always some room for manoeuvre, in every specific sport practice, from professional soccer to recreational cycling. In every case there is a constant hovering between the dominant trend (sport as an instrumental means to utilitarian ends, say power or success) and the recessive trend (a rich conception of sport as a full-fledged, *eudaimonistic* being-in-the-world).

The *poion-poson* distinction outlines the polarity of human sportsmanship. *Poion* and *poson* are complementary dimensions of sportive phenomena. This polarity defines a continuum on which individual athletes can be scored as it were. Seven times Tour de France winner (1999–2005) Lance Armstrong, for instance, although characterised by Van den Bossche as a posonistic, calculative rider, also bears poionistic traits. Notwithstanding his focus on quantity, there is a hint, even in Armstrong (nicknamed *The Boss* or *King Lance*) of a hip romantic,[36] a post-Nietzschean qualitative holist who re-valued all the existing cycling values. As a straightforward, 'square' Texan, Armstrong challenged the rusty European standards of cycling that were taken for granted until then. He, for instance, turned the time trial into something completely different, by introducing high cadence spinning (*souplesse*!) instead of using the big gear (brute force!). He also frequently

[36] Armstrong can best be seen as a late-romantic, a former *Sturm und Drang*-adherent which has come of age.

challenged the tacit mores of the peloton. He trusted on his long training sessions, submerged in Van den Bergian solitude, and concentrating on the *Tour de France*, the *Sancta Sanctorum* in professional cycling. "Armstrong is a rarity because he prefers to go out on his bicycle day in and day out, rather than upgrade his form in races" (Abt 2001).[37] In this sense, he rather personifies Pirsig's John Sutherland, who simply rode and trusted his intuition.

As a cancer survivor and cancer fighter, leading his Livestrong-foundation and his homonymous team, Armstrong furthermore was a man with an unmistakable mission; a role model with a high societal impact. What of course tipped the *poion*-scales the other way again was his démasqué as a consistent doping sinner in 2012. Still, counter-balance can be found in Armstrong's return to his first love in sport, the triathlon, after ending his professional cycling career in 2011, a remarkable return to the preferred poionistic attitude again, which never became completely extinct. By pushing the horizon of human possibility to the limit, this combination of long distance swimming, cycling and running, the classic triathlon is the ultimate manifestation of the ascetological-ecosophical attitude, as outlined in the previous chapter.[38]

The same applies at the other side of the spectrum. Alexander Vinoukourov, according to Van den Bossche the poionistic and romantic rider *par excellence*, because he was always attacking and never calculating, turned out to be a hell of a fraud.[39] He was unmasked as a dope-sinner and he allegedly won the 2010 edition of the cycling classic Liège-Bastogne-Liège only by bribing Russian Alexander Kolobnev, who broke away from the pack with Vino at the *Roche aux Faucons*, at 16 kilometres before the finish line. Kolobnev was said to have let the Kazakh win in a half-hearted sprint after a short private chat in the last kilometre.[40] When confronted with this, Vinoukourov did not deny that a payment was made to Kolobnev, but he said that he did not really buy the race: "It's my private life. I often loan money left and right" (Velo News).[41]

---

[37] Lance Amstrong in the *New York Times*: "You have to have a basic gift and then it's how you work with that gift, how you shape it, the work that you do, the intensity you do it in and then the motivation for the race. I'm very motivated for this race. It's everything" (Cited in Abt 2001).

[38] This critical point will be elaborated in Chap. 7*: On Agon and Ecosophical Endurance: Finding your Own Pace*. By revaluing the Sisyphean concept of *askēsis* as a contemporary and relatively clean and green means to fight the struggle with and against oneself, we may overcome the uncompromising fight against our natural environment and transform the human condition into a 'coexistent agony' wíth the world. This, however, without throwing the competitive, proto-agonistic baby that still lingers in us out with the bathwater. The classic triathlon embraces both impulses: coexistence ánd competitive agonistic strivings.

[39] Van den Bossche was not aware of Vinoukourov's treacherous ways or Armstrong's doping usage when he wrote his cycling book.

[40] It has to be noticed that this is a common practice in professional cycling as of the early days. Cycling has always been a game of haggling, covered by a peloton-wide *omertà*, a code of silence. The contemporary high-tech registration of races has just made this more visible.

[41] Vinoukourov and Kolobnev originally had to face judgement in a Belgian court on March 13, 2018. The lawsuit was suspended, however.

The critical reversal of the Armstrong-Vinoukourov case learns that the quality-quantity watershed between 1974 and 2010 is not a matter of either/or, because both dimensions are always present. The bottom-line of Van den Berg's 'doctrine of change', applied to sport, is that every epoch has its typical dynamics. Synchronous facts that draw the attention of the beholder *must* hang together, "but also other, unmentioned but synchronous facts in principle comply with the law of synchrony, which claims that the events at hand are related in a significant way and form a pattern, an *image*" (Zwart 2002, p. 169). The *poion-poson* watershed allows us to explain the contrast between the image of 1974 and the image of 2010, and therein resides the added value of metabletics.

The metabletic comparison also allows us to evaluate changes in the norms and standards of sports, since these are always human activities embedded in fluctuating historic settings. Metabletics helps us to determine the moral value of a remarkable sportive event, regardless of the specific motives of the athlete at the time. In other words, metabletics claims to understand the athletes better than they understand themselves. The metabletic reading allows us to confer meaning to concrete sporting acts and to explain why they continue to appeal to us. The *explanandum*, the phenomenon itself (say the allegations of doping or bribing in professional cycling), and the *explanans* (the worded explanations of that same phenomenon by the athletes and coaches involved) may diverse, but from a metabletic perspective there is something posonistic about in doping cases in 2010 and something poionistic about those in 1974.

The resulting subjective 'thick descriptions', a phrase introduced in anthropology by Clifford Geertz, but originating in William James (Toulmin and Gustavsen 1996, p. 209), are a necessary condition for the scientific study of human sportive behaviour, which has to deal with *Zeitgeist* fluctuations. Understanding human behaviour requires a broader view, a thick and creative use of language. "The thick descriptions cultural anthropologists give of the activities they study are not written in the dry, cool, abstract language of physical theory, but in the rich, concrete, colourful language of daily life" (p. 209). According to Stephen Toulmin, writings of 'the anthropological kind' (which include the broad internalist take on sport, I argue)[42] therefore rather "resemble reports by investigative journalists more than they do the papers of technically trained physicists" (p. 209).

Van den Berg's style of writing can also to be characterised as 'thick'. He not only thoughtfully identifies *prima facie* randomly chosen but on closer inspection metabletically coherent facts, representing moments of transition in history, he also looks at the incidents and anecdotes, the backstage of the concerning era. By means of his imaginative casuistry and meaningful asides, he follows the journalistic adage 'show, don't tell'. Moreover, his elliptic and apodictic sentences address the reader directly with an unequivocal message: leave the levelled path that leads to doom, contemporaries.

---

[42] In the next chapter, *Ascetic Practices, Hermeneutical Cycles and Ecosophical Endurance*, I will elaborate on the comprehensive style of broad internalism, which comes quite close to hermeneutics and phenomenology, I argue.

Whilst questionable from a strict historiographic or social scientific[43] perspective, the metabletic method has its benefits for tentatively approaching the practical-philosophical subject-matter of this study: How can endurance sport at large and cycling in particular contribute not only to self-knowledge, but also to self-improvement and to sustainability? Looking through metabletic lenses, and armed with the dichotomy of *poion* and *poson*, I concluded that Van den Berg's fear of a scorching spinalism definitely applies to 2010, but also allows us to explain what was so exceptional about 1974.

Furthermore I have attempted to show that multiple perspectives are possible. What at first sight looks like sheer quality may harbour calculative traits as well, and vice versa. This proviso does not undermine the explanatory power of Van den Berg's metabletic disapproval of the reflex as such. What he seems to have overlooked (or willingly ignored), however, is that 'acting over the spine' as a result of technological progress also provides unique opportunities for finding real life. In other words: *poson* enables *poion*. Instead of melancholically focussing on supposed original perceptions of earlier, pre-technological times,[44] and supposedly more reflective days, we should also acknowledge the potential benefits of technology for a life lived to the full. Again, we should look *through* the glasses, not just *at* them.

Now the remaining question is: how can a critical assessment of (sport-)technology provide an escape route for the stalemate between idealistic ponderings and concrete scenarios for a sustainable concept of endurance sport?

4.3 Taking up Thread of the Ascetological–Ecosophical Perspective

Van den Berg interprets the massive turn to technology as a symptom of estrangement from being truly human, fully aware of the larger scheme of things at every single step. Particularly when it comes to modern ways of human transportation Van den Berg is implacable: automated movement practices result in spinal behaviour. The relation is causal, not correlative. This negative interpretation of technology as un-escapable estrangement *per se* concurs with Peter Sloterdijk's idea of *allo*-technology (*allo* means 'other' or 'alien' in Ancient Greek), already touched on in Chap. 1. This highly alienating type of technology bears the characteristics of a calculative, *posonistic* attitude and confirms Van den Berg's spinal intuition.

[43] At least in the perspective of the data-driven empirical turn that the social sciences have undergone during the last decades.

[44] Times that according to Lemmens never have existed, since mankind is technological per se. "[T]he human condition as such *is*—that is to say is nothing but or boils down to—the technological condition. Man is the effect of the becoming-technical of life" (2008, p. 521).

Allo-technology diverts us from our existential destination, which is: being an observant practitioner, carefully considering every single step he or she is about to take.

Sloterdijk, however, also provides an alternative and more inviting perspective on technology and its inherent tendency towards an automated, reflexive life. His idea of human-friendly and nature-friendly *homeo*-technology, also already briefly presented in Chap. 1, attempts to fold itself to the curved and winding ways of nature rather than trying to tame or rectify them. This is more the co-evolutionary type of technology, which takes qualitative aspects into account and strives for a manageable coexistence between humans and other species. One could think of less polluting ways of transportation, or introducing types of skis that are less damaging to ecologically vulnerable areas. As such, *homeo*-technology potentially carries a germ of *poion*, an impulse for critical human reflection on the quality of a life that is not just running over the spine, but also reflects quality and meaning.

Nevertheless, also in *homeo*-technology, the question remains: how to find the right mean between an increasingly automated world and the possibility for conscious musings off the beaten track? How to regain a sense of true and un-mechanic physicality as a tool for excavating and expanding the ascetological-ecosophical self?

As outlined, in *You must Change your Life: On Anthropotechnics* (2009/2013) Sloterdijk contends that the call for *metanoia*—a turning point in life, a radical personal reform—has been at work since time immemorial, but manifests itself differently in different eras. Over the last decades the athletic ideal, which was so predominant in ancient Greece, has become very influential once again. Sloterdijk tenaciously emphasizes that we seem to have forgotten, however, that it takes a lot of *askēsis* to become an athlete. The good life doesn't come without a substantial effort.

When it comes to 'movement practices'—from running and cycling to gymnastics—it takes a lot of practice to reach a state of 'natural' smoothness. Sigmund Loland's ecosophy of sport (1996) seems especially viable for thinking about sports that require highly developed motor skills, such as skiing or various sorts of ball-games. Here diligent training practices are necessary conditions for condensed outburst of hilarious joy during races or games. Dull preparatory activities ideally result in rich ends. In Sloterdijk's all-embracing ascetological perspective, training and games or contests are not just means to ends—diligent practising linearly leading to eudaimonistic joy—but stopovers on the eternal spiral of quasi-Sisyphean human exertion. Low intensity training and high intensity racing are two sides of the same coin: they both reflect the human ascetic predisposition.

Especially those who are able to push their limits the most, preferably with a grimace on their face, deserve our esteem, Sloterdijk contends. This propensity for the suffering explains his preference for endurance sport. For him from a motor point of view highly repetitive, or even monomaniac, road cycling is the ultimate modern variant of ancient heroism. Particularly the mountain stage, where the dull and thoughtless horizontality (spinalism!) of everyday life is firmly challenged by *Vertikalspannung*, our inborn tension for vertical challenges makes us stand in awe. Those who climb forbidding mountains embody true and pure ascetic existence. This is why *homo repetitivus* (repeating man) or *homo artisa* (training man) appeals to our imagination at the utmost. Even more: we all want to become an ascetic of our own, and the bicycle is the perfect tool for realising this vertical imperative.

Sloterdijk's ascetology cycling enables us to overcome the dichotomy of *homo compensator* and high-end technology in a magnanimous and an up-to-date way. A way that allows us to attenuate our energy and resource-consuming lifestyle without forcing us to go back to a simple life of staying at home. Almost everyone can ride on the flat, downwind in the peloton, but only those who prove capable of fighting a duel on the steepest mountain slopes are the real heroes.

But before climbing mountains you must learn to overcome the initial fear of falling and to properly handle a bike. Once you're able to overcome the gravitational challenges that come with cycling (and just have learned *not* to be aware of every single pedal-stroke!), you must train a lot to become a full cyclist, probably on the flat and in a group. Van den Berg's disapproving quote in which he contends that "(t)he reflex discerns herb nor tree, knows no nature; the reflex has no land, no homeland" (1973, p. 117) is actually a subscript next to a picture of small peloton of 10 riders. They are cycling with a tailwind[45] in a typical Dutch landscape. They harmoniously cycle on a dyke. At their right hand side there is a river or canal with reeds on the shore. At their left there is a row of poplars and a small wooden house. For good reasons the riders only seem to concentrate on the back wheel of the rider in front, since if they wouldn't, they'd probably crash.

Because of this total (and other than the pedalling itself highly conscious!) fixation on the spinning back-wheel of the rider just in front, they do not seem to be preoccupied with landscape musings. We still should ask ourselves: is this type of cycling, loathed by Van den Berg's, totally spinal? Or are these cyclists rather in a 'flow', to use a term often employed to depict a state where mind and body are said to merge into a state in which the sporting subject seems to be detached from the concrete world? Are these cyclists subconsciously preparing for a *démarrage*, a sudden jump out of the peloton, to arrive solo at the finish-line? In other words, aren't also Van den Berg's cyclists combining reflexivity with reflection, giving rise to a situation which could be referred to as reflective spinalism?

---

[45] An interesting detail, since Dutch cycling pictures in newspapers and magazines often show heroic riders beating a strong headwind. Probably because this gives a more dramatic effect. This image is epitomized in a famous song by national bard Boudewijn de Groot, entitled *De eenzame fietser* (*The lonely cyclist* 1973). "How strong is the lonely cyclist who, bent over his handle bar, fighting a headwind, finds himself a way?" (My translation).

Geschiedenis van de revoluties, geschiedenis van de reflex

Helpt de gevonden formule beter begrijpen waarom tussen de geschiedenis van de revoluties en de geschiedenis van de reflexleer de overeenstemming van de jaartallen bestaat? Zeker is nog veel duister. Maar het begin van een opheldering is niettemin gegeven. De lezer lette opnieuw op de term *moi commun*. Om dit *algemene ik* gaat het, in alle vier revoluties. Van dit *algemene ik* mag niemand verwachten dat het een sonnet schrijft. Noch dat het drie, vier talen spreekt. Noch dat het een muziekinstrument beheerst. Dat soort spierbewegingen is aan het enkele ik voorbehouden. Het *moi commun* krabt zich, op de plek waar de vlo bijt. Het *moi commun* strekt het been, als dit been een tik onder de knieschijf krijgt. De reflex is de beweging van allen. De reflex kent zelfs landsgrenzen niet. Niet de arbeider is het, die geen vaderland bezit, de reflex heeft geen vaderland.
Het verband tussen de geschiedenis van de revoluties en de geschiedenis van de reflexleer is zo duister niet meer. Maar wel lijkt het nodig dit verband, dat nog slechts met enkele zinnen staat aangeduid, meer inhoud te geven. Tot dat doel dient het volgende hoofdstuk.

117

De reflex ziet boom noch kruid, kent geen natuur; de reflex heeft geen land, geen vaderland

Challenging Van den Berg's spinal picture. Subscript next to the picture: "The reflex discerns herb nor tree, knows no nature; the reflex has no land, no homeland". (Source: *De Reflex: Metabletische tegelijk maatschappijkritische studie* (1973))

Perhaps these spinal cycling machines are about to enter a state of super-normal perception, referred to by sport philosophers Douglas Hochstetler and Peter M. Hopsicker as "the heights of humanity":

> (W)e contend that endurance sport is notably different from other movement forms in terms of available experiential quality and, as such, is emblematic of the depths of experience available through commitment to sporting practices (2012, p. 118).

Reflexivity may provide unique opportunities for experiencing the depth of humanity "in the strenuous vein" of the committed endurance athlete (p. 132),[46] in accordance with pro-cyclist Peter Winnen's observation that although during his ascents of the Alpe d'Huez in the early eighties his senses were narrowed down to mere survival, the gruesome French mountain nevertheless entered his body in an unconscious, subliminal way (2000, p. 234). The Alpe somehow engrained itself in Winnen's muscle memory.

[46] Runners and riders rather than joggers and so-called 'weekend warriors', who ride their bicycles on (sunny) Saturdays or Sundays only.

Elite athletes are able to act sub-consciously, or '*Empraktisch*', to use a neologism coined by the German philosopher Volker Caysa (2008). This is a form of immediate rationality and functionality without conscious rational experience. One knows what one does, reflexively. This empractical experience of movement practice, in which the higher brain centres are not actively involved, equals Van den Berg's spinalism. But still: one somehow really knows. As Caysa justly argues: a sportsman or –women doesn't have to know the details of physics, biomechanics or neurobiology to know how to operate in a functional way during challenging sport situations.

What makes Johan Cruyff such an exceptional, hyper-conscious, 'empractical' player and Merckx a skilled and daring cyclist (still able to reflect), is exactly that they are capable to act 'rationally via the spine'. Cruyff on many occasions has stressed the importance of consciousness, self-control and reflectiveness acquired through training, which seems the very opposite of the primitivism of sheer reflexes and just acting over your spine. "You play football with your head, and your legs are there to help you."[47]

Caysa's concept of empracticality sheds a different light on Van den Berg's idea of spinalism. The crucial point is that these motor acts first consciously have to be engraved in the neural architecture of the body, including the conscious brain. Only after a long training process[48] these acts can become automated. Through devoted training the elite athlete is able to acquire 'empractical' skills that during their performance do not have to be explicitly remembered every single time, they may be 'forgotten'.

Caysa even argues that in sport the *ego* should become *id* again. Van den Berg, on the contrary, wants the *id* to become a highly conscious 'I' again; fully aware of every single step it takes on the bumpy and meandering road to a meanigful life. The decisive point is that Van den Berg interprets sport in general and cycling in particular as a mechanical escape from true human destiny, whereas thinkers such as Sloterdijk and Caysa try to get in the shoes of the seriously sporting subject itself. They attempt to grasp the 'hidden essentials'[49] of our ascetic predisposition.

Also the Norwegian sport philosopher Gunnar Breivik provides further sport philosophical ammunition for the idea that there is more to highly skilled rule-

[47] NCRV-gids (Dutch radio and tv-guide), June 12 1982, nr. 24, my translation).

[48] It is often argued that it takes 10,000 h to master specific practices of all sorts, form playing piano to scoring goals at elite level. (E.g. Malcolm Gladwell's thesis of the necessity of a 10,000 h investment as the *Tipping-Point* of greatness (2000)). This is an arbitrary figure, since more talented people probably need less time investment and untalented far more. In either case: it takes a lot of time to become an automat. Even more, the ability to act automated during a game under high pressure has also to be maintained after putting in 10,000 h.

[49] In quotation marks to circumvent the onto-philosophical problems associated with (hidden or overt) essentialism.

bound movement practices than just mechanic 'Zombie-like'[50] behaviour. There is room for manoeuvre between reflexion and reflection, he argues.

> But my point is that the crucial elements in elite performances have to be continuously attended, and non-conscious dealing occurs since experts need not monitor everything they do. Nevertheless, automatic dealing has previously been consciously carved out and installed in the nerve-muscle system.
>
> Phenomenal consciousness is important. In the end, the feeling of what it is like to give a good performance and win a race is one of the most phenomenal feelings one can have (2013, p. 104).

It is obvious that elite performances have special benefits to offer when it comes to fully understanding what high level sport is about. This 'phenomenal feeling' is not just a privilege of the elite. All sorts of intense movements may fill this subtle equilibrium between automation and deliberation. The resulting overwhelming sensation of joy may occur during the widest imaginable range of ascetic efforts. From an extensive long solo training ride on paved roads to a high intensity cross country mountainbike race. From a sunny Sunday afternoon ride with a group of riders of mixed age and ability to a merciless championship race. Automatic movements may harbour a spark of *eudaimonia*; an ascetologically enhanced holistic experience of ecosophical joy. Pushing Van den Berg's disapproving spinal intuitions into a more positive direction results in the recognition that semi-automated serious cycling can be the very antidote to a life just running over the spine. Being able to spin enables you to think things over. Reflexion and reflection are complementary dimensions. And this explains why cycling furthers life.

The re-enactment of reflection in reflexion on two un-motorised wheels in a light-weight frame might even lead to *metanoia*. This radical turn for the better in life overcomes the dichotomy between reflexive quantity and reflective quality. Knowing how your bicycle works, Knowing its fine mechanics, learning to handle it properly and (over time) thoughtlessly, riding alone or in a group, uphill or on the flat. These are the prerequisites to truly enjoy the strenuous and enduring life in all its modalities. Even under highly competitive circumstances during which riders indeed seem to discern neither herb nor tree, nor to know nature, they still may enjoy an occasional touch of the good life. The reflex is a necessary condition for truly experiencing a meaningful, reflective existence and paves the road towards the ecosophical homeland.

---

[50] Running up Zombie hill at the final part of the Norseman extreme triathlon (Cfr. Chap. 1 *Prologue: The Good Life, Endurance and Sustainable Cycling*) is definitely more than just Zombie-like behaviour, I argue. While the remaining contestants are not attentive at every single step anymore, they still have to be highly alert.

References

Abt, S. 2001. Training, not racing, gives Armstrong his edge. *New York Times*. http://www.nytimes.com/2001/07/29/sports/cycling-training-not-racing-gives-armstrong-his-edge.html. Accessed 25 Jan 2018.

Bertz, E. 1900/2012. *Philosophie des Fahrrads*. (Erweiterte Neuausgabe herausgegeben von Wulfhard Stahl). Hildesheim: Georg Olms Verlag.

Breivik, G. 2013. Zombie-like or superconscious? A phenomenological and conceptual analysis of consciousness in elite sport. *Journal of the Philosophy of Sport* 40 (1): 85–107.

Brown, S. Undated. *Standing while cycling*. https://www.sheldonbrown.com/standing.html. Accessed 31 Jan 2018.

Caysa, V. 2008. Körperliche Erkenntnis als Empraktische Körpererinnerung. In *Körperliche Erkenntnis – Formen reflexiver Erfahrung*, ed. F. Bockrath, B. Boschert, and E. Franke, 73–89. Bielefeld: Transcript Verlag.

Chen, S., W. Kim, S.M. Henning, C.L. Carpenter, and Z. Li. 2010. Arginine and antioxidant supplement on performance in elderly male cyclists: A randomized controlled trial. *Journal of the International Society for Sports Nutrition*. https://doi.org/10.1186/1550-2783-7-13.

Close, J.-P. e.a. 2016. *AiREAS: Sustainocracy for a healthy city: The invisible made visible phase 1*. Cham: Springer.

Cycling Greats. *Eddy Merckx*. www.Crazyaboutbelgium.co.uk. Accessed 17 July 2010.

Drabinski, J. 1993. Radical empiricism and phenomenology: Philosophy and the pure stuff of experience. *Journal of Special Philosophy, New Series* 7 (3): 226–242.

Gladwell, M. 2000. *The tipping point. How little things can make a big difference*. New York: Little, Brown & Company.

Gorris, J., and D. Kurbjuweit. 2008. Philosopher Peter Sloterdijk on the Tour de France: 'The riders are just regular employees'. *Der Spiegel Online International,* 10 July: http://www.spiegel.de/international/europe/philosopher-peter-sloterdijk-on-the-tour-de-france-the-riders-are-just-regular-employees-a-565111.html. Accessed 17 Nov 2017.

Hilvoorde, I., J. Vorstenbosch, and I. Devisch. 2010. Philosophy of sport in Belgium and the Netherlands: History and characteristics. *Journal of the Philosophy of Sport* 37 (2): 225–237.

Hochstetler, D., and P. Hopsicker. 2012. The heights of humanity: Endurance sport and the strenuous mood. *Journal of the Philosophy of Sport* 39 (1): 117–136.

Humanity + Transhumanist declaration. http://humanityplus.org/philosophy/transhumanist-declaration. Accessed 10 Jan 2018

Husserl, E. 1936/1970. *The crisis of European sciences and transcendental phenomenology*. Evanston: Northwestern University Press.

Illich, I. 1974. *Energy & equity*. New York: Harper & Collins.

Johnstone, J. Undated. Triathlon – The early history of the sport. http://www.triathlonhistory.com/. Accessed 18 Dec 2017

Kok, A. 2004. *1974: Wij waren de besten*. Amsterdam: Thomas Rap.

Krechtmar, S.R. 1974. Modes of philosophic inquiry and sport. *Journal of the Philosophy of Sport* 1: 129–131.

Landeweerd, L. 2016. *Das verborgene Leben der Dinge/The hidden life of things*. Unpublished Lecture/ artistic object.

Loland, S. 1996. Outline of an ecosophy of sport. *Journal of the Philosophy of Sport* 23: 70–90.

———. 2001. Technology in sport – Three ideal-typical views and their implications. *European Journal of Sport Science* 2 (1): 1–11.

Long, E.T. 2003. *Twentieth-century western philosophy of religion*. Dordrecht: Kluwer Academic Publishers.

Mook, B. 2009. The metabletic method: An interdisciplinary look at human experience. *Phenomenology & Practice* 3 (1): 26–34.

Mumford, S. 2014. The aesthetics of sport. In *Bloomsbury companion to the philosophy of sport*, ed. C. Torres, 180–195. London: Bloomsbury.

Nietzsche, F.W. 1954. *Die fröhliche Wissenschaft*. München: Werke in drei Bänden.

———. 1967. *The will to power*. New York: Vintage Books https://archive.org/stream/FriedrichNietzscheTheWillToPower/Friedrich%20Nietzsche%20-%20The%20Will%20to%20Power_djvu.txt. Acessed 5 Feb 2018.

Pirsig, R.M. 1974. *Zen and the art of motorcycle maintenance: An inquiry into values*. New York: William Morrow.

Plato. 1981. *The republic*. Harmondsworth: Penguin.

Reid, H.L. 2017. Why Olympia matters for modern sport. *Journal of the Philosophy of Sport* 44 (2): 159–173.

Rousseau, J.-J. 2012. *Du contrat social, ou principles du droit politique*. Collection complète des oeuvres, Genève, 1780–1789, vol. 1, in-4°. Édition en ligne www.rousseauonline.ch. Accessed 16 Jan 2018.

Sloterdijk, P. 2009/2013. *You must change your life – On anthropotechnics*. Cambridge: Malden.

Stiegler, B. 2014. *The anthropocene and neganthropology*. Canterbury. https://www.academia.edu/12693668/Bernard_Stiegler_The_Anthropocene_and_Neganthropology_2014_. Accessed 12 Dec 2017.

Suits, B. 1978/2005. *The grasshopper: Games, life and utopia*. Ontario: Broadview Press.

Torres, C.R. 2012. Furthering interpretivism's integrity: Bringing together ethics and aesthetics. *Journal of the Philosophy of Sport 39*: 299–319.

Toulmin, S., and B. Gustavsen. 1996. *Beyond theory. Changing organizations through participation*. Amsterdam/Philadelphia: John Benjamins Publishing Company.

Van den Berg, J.H. 1973. *De reflex. Metabletische, tegelijk maatschappijkritische studie*. Nijkerk: Callenbach.

———. 1996. *Geen toeval. Metabletica en geschiedschrijving*. Kampen: Kok Agora.

Van den Bossche, M. 2005. *De passie van Marc Van den Bossche: Wielrennen*. Rotterdam: Lemniscaat.

Velo News. http://www.cyclingnews.com/news/vinokourov-and-kolobnev-still-facing-possible-corruption-trial/. Accessed 25 Jan 2017.

Winnen, P. 2000. *Van Santander naar Santander – Brieven uit het peloton*. Amsterdam: Thomas Rap.

Winner, D. 2000. *Brilliant Orange – The neurotic genius of Dutch football*. London: Bloomsbury.

Zwart, H. 2000. A short history of food ethics. *Journal of Agricultural and Environmental Ethics 12*: 113–126.

———. 2002. *Boude bewoordingen – De historische fenomenologie ('metabletica') van Jan Hendrik van den Berg*. Kampen: Klement.

Chapter 5
Ascetic Practices, Hermeneutical Cycles and Ecosophical Endurance

By producing new configurations between contemplation and fitness, the current 'renaissance' enables new festivals on the mountain of improbabilities

(Peter Sloterdijk 2009/2013, p. 155).

Abstract After discussing the environmental effects of sport practices and putting a metabletic view on sport to the test, I now will turn to philosophy of sport as an academic discipline. Like other branches of applied philosophy, such as animal ethics and environmental philosophy, it started off in the early seventies of the previous century. Currently, it seems locked in a binary view. On the one hand narrow internalists, or formalists, argue that sports are uniquely constituted by their rules. This point of view can be referred to as the *autotelic* stance. Herein sport is considered an end in itself, constitued by rules and (eventually) shared conventions on how to play or race well. Broad internalists, or interpretivists, on the other hand, contend that sport is more than just a gratuitous and playful end in itself. In this line of reasoning sport also can be a means toward other ends: national pride, prize money, a ruthless quest for records, challenging the existing order or advancing international peace. This is the *heterotelic* view.

In this chapter I will revitalise the reflection on sport as a dimension of the human condition by attempting to move beyond the binary opposition of internalism and externalism. I will do so by focussing on the potentially positive aspects of the concept of *agon*, a term which denotes struggle or contest. As an 'agonal' or competitive social practice, sport turns out to be a means to an end, in the sense that it surpasses the concept of sport as self-referential play: seeking knowledge, understanding the human condition, and cultivating virtue. I argue that this agonistic heterotelic view seems the better option.

In order to strengthen my claims I will uptake, broaden and deepen Peter Sloterdijk's ascetology already introduced in previous chapters. The bottom line of his call for a change for the better is that we have to become aware of the fact that our 'ascetic planet' is inhabited by individuals who are constantly and relentlessly

R. Welters, *Towards a Sustainable Philosophy of Endurance Sport*, Library of Ethics and Applied Philosophy 37,
https://doi.org/10.1007/978-3-030-05294-2_5

training themselves. This may be self-focused, but it may also have a broader scope: we train ourselves to become better humans, contributing to a just and sustainable society. Paradoxically, however, this will only work when we become aware of our exercises as forms of life that engage the whole practicing person.

A broad internal hermeneutic interpretation and furthering of endurance sport, especially cycling, can enrich our understanding of this sports activity as a form of asceticism. By following and furthering this ascetological imperative we can elaborate a view on cycling as an upwardly oriented 'spiral' that can contribute not only to self-knowledge and self-improvement on the individual level (*metanoia*), but also to an 'ecosophical renaissance' on the collective level.

Keywords Autotelic view · Heterotelic view · Ascetology · Hermeneutic cycles

5.1 Playing Games

Since its rise in the early seventies of the twentieth century[1] most of the intellectual energy in the academic philosophy of sport has been spent on defining 'sport',[2] or the broader category of '(physical) games'.[3] Due to its analytic, 'Anglo-saxon'[4] roots this still flourishing branch of applied philosophy can basically be seen as an attempt to refute Ludwig Wittgenstein's notorious (and among sport philosophers often quoted) lament on the difficulty of defining activities that in common speech are denoted as'games'.

> Consider for example the proceedings that we call 'games'. I mean board-games, card-games, ball-games, Olympic games, and so on. What is common to them all?—Don't say: 'There *must* be something in common, or they would not be called "games"'—but *look and see* whether there is anything common to all (1953/1978, p. 31).

Take for instance a closer look at ball-games, Wittgenstein suggests. This is a serious matter of winning or losing in the case of football, but nothing but a playful and obligation-free activity when a child throws a ball against the wall and catches it. Since they do not share a common essential ground, he argues, games can only be characterized in terms of 'family resemblances'. There are multiple similarites, but

[1] See for a detailed account of the genesis of sport philosophy Lunt and Dyreson (2014, p. 30).

[2] A noun etymologically referring to playful or 'ludic' physical activity distracting from serious human duties, but as of the nineteenth century also referring to "competitive physical activities" (Fry 2014, p. 371).

[3] The rather confusing use of 'sport' and 'games' is a conceptual obscurity which still slumbers in the philosophy of sport, and which to some extent has to be attributed to the different connotations of play, games and sport in different languages. In North America sport is usually strongly connected with professionalism, a strong sense competiveness and commercialism, whilst in Europe traditionally rather is associated with 'gentle(wo)manlike' amateurism, literally: the love for sport. This subtle difference still is silently echoed in many sport philosophical publications.

[4] Cfr. footnote 148 for a refinement of this generalisation.

no compelling and overarching formal structure discernible all these activities we call 'sport' or 'games', Wittgenstein reasons.

In *The Grasshopper: Games, Life and Utopia* (1978/2005) the Canadian sport philosopher Bernard Suits takes up the gauntlet, and offers a witty retort to Wittgenstein's call for a covering definition of games beyond mere similarities. The crucial point of Suits's critique is that Wittgenstein did not follow his own order to really look and see. "He looked, to be sure, but because he had decided beforehand that games are indefinable, his look was fleeting, and he saw very little" (p. 11).

According to editor Thomas Hurka, *The Grasshopper* is "a precisely placed boot in Wittgenstein's balls" (Suits 1978/2005, introduction, p. 11). The title of the book refers to Aesop's famous fable of the grasshopper and the ant. By light-heartedly hopping around all summer long, and refusing to store up for winter, the former represents the model of providence and playfulness, while the food stocking industrious ants stand for the serious life. A thorough, consequential and witty conceptual analysis of the grasshopper's undeniable 'ludic' nature leads Suits to proposing the following definition of game-playing:

> To play a game is to attempt to achieve a specific state of affairs [prelusory goal], using only means permitted by rules [lusory means], where the rules prohibit use of more efficient in favour of less efficient means [constitutive rules], and where the rules are accepted just because they make possible such activity [lusory attitude]. I also offer the following simpler and, so to speak, more portable version of the above: playing a game is a voluntary attempt to overcome unnecessary obstacles (Suits 1978/2005, p. 54-55).[5]

Ten years after 'the Grasshopper', Suits reconsiders the differences between games, play, and *sport* in *Tricky Triad: Games, Play, and Sport* (1988). Now he contends that 'primitive play' is not focused on athletic skills, but skills may develop through the repetition of the play. He discusses the Olympic Games, professional athletes, the desire to win, and the philosophy of sport at large.

> Games and sports are enterprises, I suggest, in which the exhibition of skill is the paramount consideration, and I'm going to argue that in play the exhibition of skill is not the paramount consideration (p. 1).

In Suits's line of reasoning, sport, then, is a matter of organised and rule-bound game-playing.

In the same issue of *The Journal of the Philosophy of Sport* the German sport-philosopher Klaus Meier (1988) criticizes Suits, concentrates on the philosophy of sport, play and games and explores their significance for an individual and their functions within different societies. Meier argues that sports are differentiated from games because sports require a demonstration of *physical* skill, whereas games as such do not.

My pragmatic synthesis, which I will try to make hermeneutically viable in the course of this chapter, is that sport is a serious and agonistic, but not per se *organised competitive* physical activity. I would, for instance, qualify (individual) running, fitness and long bicycle-rides not only as health enhancing physical exercises,

---

[5] Note that this initial definition is not about sport per se, but (just) about playing (physical) games.

but under circumstances (getting better, improving your personal best, racing against imaginary opponents) also as (auto-agonistic) 'sport'. And, similar to Meier, I consider chess as a game that is played agonistically, but not as a sport, since the element of high-end physical skill as a necessary condition to play successfully[6] is evidently lacking. In chess flourishing physicality is accidental, not essential (unless the physical ability to remain concentrated for hours on end is regarded as such).

In the slipstream of this definitory debate the following sport philosophical landscape has arisen. On the far end 'extreme externalists', or advocates of the "non-theory of athletic performance" as Loland phrases it (2003, p. 4), argue that sport is nothing but a mean towards other ends (power, money, nationalism), and therefore lacks a common denominator. Although at a glance actual elite sport practices (e.g. the Olympic Games or the FIFA World Cup) seem to confirm this cynical outlook on sport as the ultimate affirmation externalism—money rules the waves—the majority of sport philosophers still argue that the activities we tend to call sport do have intrinsic values. This quest for the deeper meaning of sport, or 'internalism', needs further consideration.

First it has to be noted that there is a dispute between a rigid and a flexibel vision on internalism. This res ults in a binary view on sport as either stricly autotelic or (at least partially) heterotelic. On the one hand narrow internalists, or formalists, still follow Suits by reasoning that a game is solely defined by its rules. Broad internalists, or interpretivists, on the other hand, contend that there is an intrinsic ethos in sport that reaches beyond sheer rules.[7] There is also something elusive, but still worth pursuing, referred to as "the spirit of sport" (McNamee 2015).

Over the years the latter, more 'continentally' inspired[8] ideas on how to *understand* rather than *explain* the wide range of human activities we call 'sport' have

---

[6] Rather than an accidental requirement to be able to move the pieces on the chessboard. The end-result of a game of chess is not determined by the physical quality of the moves but by their strategic result. A trembling hesitating hand that results in check-mate is better than a powerful and determined move that leads to the loss of an advancing pawn.

[7] For a more detailed account, which also discusses conventionalism with its anti-formalism and its focus on the (tacit) ethos of games, cfr. Lopez Frias (2014), who argues that William J. Morgan's deep conventionalism "could be defined as an internalist type of conventionalism.... For Morgan, deep conventions constitute sports' normative inner nature. These conventions have been historically constructed and accepted by practice communities through careful deliberation. On the other hand, coordinating conventions are arbitrary agreements that serve to solve practical problems (they could be grounded in elements, external to sports, for instance, economic interests" (2014, p. 363).

[8] Analytic philosophy (AP) geographically is said to refer to the Anglo-American world, whereas continental philosophy (CP) usually is associated with mainland Europe. "But this distinction is very limited if not inaccurate. Both CP and AP emerged out of the same philosophical tradition as the work of Kant, Bernard Bolzano and Franz Brentano. All three were from the European continent. Furthermore, there are quite a few philosophers living in mainland Europe who are mostly concerned with AP, as there are a number of philosophers in North American universities who are mainly concerned with CP. So, in the modern world, geography indicates little about philosophical preferences. (…) AP has shared the problem-oriented and scientific-empirical outlook of the natural sciences, that is, its focus is to search for knowledge and truth in a rationalistic framework where a problem becomes divided and reduced into its smaller parts, analysed and explained in

gained more influence (Fusche Moe 2014, p. 53). Approximately since Alisdair MacIntyre's *After Virtue* (2007/1981)—with its fertile distinction between formal institutions and their external goods and vivid practices driven by internal goods—continentally rooted hermeneutics, or interpretivism,[9] has entered the sport philosophical stage. Javier Lopez Frias (2014) provides a helicopter view on the current situation in sport philosophy. It results in a reluctant inclusion of some elements of externalism, as a perhaps unfortunate but, nonetheless, inevitable ingredient in contemporary money- and prestige-driven elite sport practices.

> So, the philosophy of sport moved from the narrow analysis of rules and social conventions to a much wider analysis that encompasses elements such as the constitutive abilities of sports, their internal goods, their models of excellence and their gratuitous logic. …It could be claimed that such a widening process still is ongoing. As a consequence, the philosophy of sport is facing a new challenge: the blurring of the limits drawn between internalism and externalism. Since sports are understood as a social practice, they are already in permanent relationship with other social practices. Should not they all share some minimum common elements? (p. 364).

In the ongoing quest for an integrative interpretive framework also Andrew Edgar, editor-in-chief of *Sport, Ethics and Philosophy* and president of the *European Association for the Philosophy of Sport*, has argued for considering sport as a serious and meaningful activity, rather than as an interesting, but in the end 'gratuitous' and self-referential philosophical riddle. Edgar reasons that there is something essential to sports that reaches beyond nominal family resemblances. Sport is not just about certain rule-bound leisure activities we happen to call sport, it also edifies a world, "albeit at the level of embodied action"(2013, p. 145). This sensitive 'aboutness' of sport calls for a continual interpretation and contextualisation – look and see!

Edgar refers to the famous boxing-match Max Schmeling versus Joe Louis in the 1930s, which "may be understood, in some sense, to be about the rise of the Nazis and a potential conflict between authoritarian Germany and the democratic United States" (Edgar 2013, p. 140). One can think of many other examples, great and small. Take the devastating 7-1 win in the world cup semi-final of Germany over Brazil in 2014 as a trans-valuation of all existing cultural values and prejudices. A match during which the Germans finally shook their chains off as hard and dutiful workers of the post-war *Wirtschaftswunder*[10] and suddenly revealed themselves as frivolous practitioners of the *jogo bonito*, while the supposed agile Brazilians were reduced to baffled bystanders. Moreover, also during an amateur intercollegiate handball or baseball-match 'prestige', 'culture', 'history' and 'hegemony' to some extent always seem to be involved. One might even dare to think of a simple solo

---

terms of logic or by the laws of nature. On the other hand, CP has been concerned with the understanding what appears to be meaningful for a person in the sense that it has tried to comprehend life as it is lived from the first person point of view" (Fusche Moe 2014, p. 53).

[9] I will use 'hermeneutics' and 'interpretivism' as synonyms.

[10] In the previous Chapter I already argued that this is a commonplace and a caricature. One could argue that in 1974 the Germans already had a touch of sophisticated *poion* in their game, albeit rather at a cerebral level than in concrete motor actions.

training ride on a bicycle as potentially more than just a ludic end in itself: getting fitter; improving one's personal best on a specific ascent[11]; overtaking another lonely rider; or pondering over the matters of life during quasi-thoughtless pedalling. What you are and how you see yourself and the world becomes visible in the way you move your pedals.

So, sport indeed is a serious case of 'world-making'. And therefore it does make sense, beyond the jingle of winning and ever reaching for faster, higher and stronger. Even more, sport can make us aware of being frail and fallible, Edgar contends.

> Worlds are made in the embodied contest of sport, or more precisely, by allowing elements of our non-sporting worlds to sediment in the constitutive rules of a sport's form. Here, in play, those non-sporting worlds are remade. In this remaking, felt in the playing and spectating of sport, we obscurely come aware of what otherwise is taken for-granted. The metaphysical and normative problems that our worlds pose can be played out within sport. Specifically, sport disrupts the second nature of the everyday world because sport's medium is the contest, or *agon*, and in that contest one must eventually fail (Edgar 2013, p. 162) .[12]

By arguing that sport can serve as a playing field for the deeper matters of life, Edgar comes close to the attempt to extract a philosophy *from* sport that Albert Piacente (2015) has undertaken:

> [A] philosophy of sport can aid in the resolution of a number of different broad-based philosophical and social issues (p. 58).

> Inverting the traditional relation between theory and practice, where practice finds its justification in terms of a theory itself justified in some prior fashion (yielding the infinite epistemological regress philosophers know only too well), by paying attention the way sport actually works we see what values are operating and from that derive a lesson (p. 72).

Sympathizing with both Edgar's tentative hermeneutical and Piacente's more pragmatic line of sport-philosophical reasoning, I will now develop my onw diagnostics building on their arguments. What values *from* sport can prove ot be of service for *which* serious lessons for everyday life? What are, in the end, the benefits for life at large that may be drawn from a truly and deeply experienced sportive attitude?

First I will uptake and further develop the views on sport described above into a more full-fledged 'continental' hermeneutic perspective. Next I will provide a more detailed analysis of the already mentioned ascetological impetuses in Peter Sloterdijk's thinking. Selected parts of the work of this contemporary and much-debated German philosopher, firmly rooted in the continental tradition, and thus in

---

[11] Since the emergence of GPS-tracking many cyclists and runners put the measurable outcome (such as distance, speed, altimeters, produced Watts) of their training sessions on the website STRAVA. This enables them to (virtually) compare performances beyond the limits of a real time race.

[12] From a statistical perspective even elite sport is rather about coping with losing than winning, since only one athlete will win the Olympic 100-meter dash, and only one team will win the champions league. A common sense argument which makes slogans as 'Winning is a mindset' or '90% of winning is mental' presumptuous if not ridiculous. Cfr. for a critical assessment of the ideology of winning Caddick and Ryall (2012).

principle closer to the "broad internalist" camp, will serve as a road-mark for this upwardly oriented spiral journey. Finally, instead of just concentrating on a few exemplary cases of elite sport, I will zoom in on one single type of sport: endurance sport, which is often seen as highly repetitive and monomaniac, thus as not very 'playful'. I will specifically focus on road-cycling, a sport or 'human practice' which is also a pet topic for Sloterdijk.

5.2 Furthering the Meaning of Sport

Suits and Suitsians finally fail to fully acknowledge the idea of 'playful' but at the same time *agonal* (that is: intensely and seriously competitive) physical activity as a possible means towards other ends—say self-knowledge or moral betterment. In the end Suits reveals himself as an uncompromising adherent of the literal 'amateur' stance of the unselfish love of sport just for its own playful sake, whereas professionalism in sport in his view automatically leads to instrumentalism. This somewhat grumpy position forces him eventually to conclude that in all sport events where the urge to win has become predominant—say the modern Olympics with its relentless motto faster, higher and stronger—lack the element of un-committal play, and therefore are not really 'pure' sport (Suits 1988, p. 8–9). I argue for a comprehensive sliding scale, rather than an exclusionary binary view.

Sport philosopher Johan Steenbergen (2004) resumes the case by proposing a more fine-grained heuristic framework to signify activities we tend to call 'sport'. He sketches a continuum from (playful) games which require 'movement abilities' which do not require a strictly organized and rule-bound setting, on the one hand, up to *agonal* games of movement skills—or straightforward organized and official competitive sports—on the other. According to Steenbergen, in his rigid amateur-professional dichotomy Suits fails to acknowledge the difference between an empirical statement (a sample survey might point out that probably all Olympians will try to win a medal) and a conceptual ontological statement (*thus* the Olympics are instrumental per se).[13] *Quod non*. While probably many participants in the Olympics

[13] During the publication of the Grasshopper in 1978 the Olympics officially still were 'amateur' only games. There was a discussion, however, on the 'state-amateurism' (or hidden professionalism) of the communist and socialist participant countries. At the same time in the Netherlands there were professional cyclists who earned less than the state-guaranteed minimum wages. Still they kept proceeding exhausting themselves to the max. They apparently felt a compelling inner urge to do so. For them a meaningful life apparently was a cycling life. This revaluation of all ludic and supposedly gratuitous values as voluntary obstacles demonstrates the idiosyncratic nature of Suits's pro-amateur-dichotomy and calls for a revision of the unproblematic conjunction of playfulness, gratuitous and true love of sport, or, literally, *amateurism*. Cfr. also Loland 1995 for the ambiguous and complicated relation in De Coubertin's modern Olympic ideology between the importance of winning (*citius, altius, fortius*!), playing, the wrongly to De Coubertin attributed idea that participating in the Games is more important than winning, and his the holistic idea of *a mens fervida in corpore lacertose* (an overflowing mind in a muscular body). This complicated vitalistic stance makes Suits's amateur-professional-dichotomy also historically highly questionable.

will at least try to win a medal, we cannot exclude the idea that participants do partake in the Olympics for playful reasons only (p. 165-166).[14]

Heather L. Reid carries another aspect into the debate on how to define sport. She contends that the idea of sport as a form of un-committal, non-serious, gratuitous and autotelic play—already put forward in Johan Huizinga's seminal *Homo ludens: A Study of the Play-Element in Culture* (1955)—impicitly still seems to be predominant among many sport philosophers, whereas the 'knowledge-seeking character of sport', largely seems to be neglected.

> [T]his play-paradigm seems at odds with the modern world, which takes sports very seriously, puts them in the service of deliberative ends, and views them (or competition at least) as essential for human thriving. Indeed our modern use of sport seems to better resemble ancient Greece, where athletic contest (*agon*) served specific political and educational goals. Huizinga claims that the ancient Hellenes simply became unaware of their contests' autotelic character …; my own concern is that we moderns are becoming unaware of—or indifferent to—sport's contemporary *ends*. … What my own study of these phenomena reveals is that sport's social and educational benefits derive not from its playful character, but from its philosophical origins as a knowledge-seeking activity (Reid 2009, p. 40).

Taking this *'heterotelic'* character of sport into serious account after all seems unavoidable when we really look for communalities in what we refer to as 'sports'.

One possible way to overcome the definitional problems of the reductive narrow internalist approach is hermeneutics. Philosophical hermeneutics is a traditional method that is "based on the idea that human beings are ontologically interpretive beings who constantly create interpretations of reality" (Lopez Frias and Monfort 2015, p. 5). This effort to understand the human condition by means of ongoing processes of interpretation has a long tradition. Etymologically, hermeneutics is derived from the Greek word ἑρμηνεύω (*hermeneuō*), meaning 'translate' or 'interpret'. A somewhat apocryhal story connects hermeneutics with Hermes, the eloquent son of Zeus and interpreter and conveyer of the Godly messages to the mortals.

During the Middle Ages and the Renaissance, the hermeneutical method became a theological tool for interpreting the Bible. In the second half of the nineteenth century, however, hermeneutics shook off its religious connections and evolved into a secular and full-blown philosophical discipline with its own method and objectives. This shift was initiated by Friedrich Schleiermacher (1768–1834) and refined by Wilhelm Dilthey (1833–1911). The latter divided the *Geisteswissenschaften* (the humanities) into three levels of possible comprehension: the highly personal sphere of *experience,* followed by *expression,* which adds meaning to experience through its discursive nature, and finally *understanding ('Verstehen'*), the highest form of 'knowledge' hermeneutics may lead to.

---

[14] The following observation makes Suits's amateur-pro dichotomy even more questionable to me. Even in highly commercialized professional sport events one can easily think of situations in which a sudden glimpse of utter playfulness pops up: a stunning but non-efficient trick by a basketball or soccer-player just for the sheer beauty of it in itself instead of winning at all costs.

In *Being and time*, written in 1927, Martin Heidegger switches over from the traditional preoccupation of hermeneutics with the interpretation of texts to a non-textified, existential understanding of 'being'. In his writings, while returning to pre-Socratic philosophy, he pleads for a more authentic way of being-in-the-world as an alternative to the positivistic-scientific way of explaining the world. Since humankind is oblivious of what being really is about (*'Seinsvergessenheit'*), we are desperately in need of a 'new titanic battle around being' (Heidegger 1927/2008, p. 21). In this campaign he seeks for αλήθεια (*aletheia*), the ancient Greek term for 'truth', although Heidegger emphasises its literal meaning: 'un-concealedness' (*Unverborgenheit*),[15] a neologism which suits his anti-positivistic objectives.

Other than transcendental philosophers, who focus on the fundamental structures of knowledge,[16] Heidegger undertakes an ontological U-turn to our experienced *Lebenswelt* (lifeworld). He develops a more authentic meaning of everyday life (*Dasein,* being there) by means of *Auslegung* (interpretation, elucidation).

> In line with Heidegger, we claim that philosophy should not aim to reduce experience by means of definition,[17]but rather it should be understood as a process of unveiling reality by embracing our experience of it (Lopez Frias and Monfort 2015, p. 10).

This iteratively interpreting, instead of narrowing down by meticulously defining, opens up a deeper understanding of sport activities beyond the infertile *autotelism-heterotelism* dichotomy.

5.3 Kynical Asceticism

In his *Critique of Cynical Reason*[18] Sloterdijk develops an alternative to the contemporary inclination towards societal cynicism, which is also discernible in some extreme forms of sport externalism. Since cynicism inoculates itself against critical resistance, Sloterdijk considers it a case of "enlightened false consciousness"(1983/1987, p. 19). In order to counter this defeatist and counter-productive attitude, Sloterdijk calls for a more authentic, literally'doggish' *cynicism*,[19] as exemplified by the ancient 'kynic' Diogenes of Synope. This down-to-earth

[15] Heidegger had a habit of converting verbs and adjectives into often dire neologisms (approx. 150 in *Being and Time*), which in German to some extent is tolerated (at least as so-called *Substantivierte Adjektive,* or 'substantived' adjectives) but in English is not. This difference in linguistic attitude is an interesting specimen of the supposed continental-analytical watershed, I argue. It reflects the broadening perspectives towards the experience of wholeness that is paramount in broad internalism versus the attempt to narrow down concepts of formalism.

[16] E.g. Kant's synthetic a priori judgments.

[17] Which, for instance, appears to be the case in Suits's before-mentioned failure to acknowledge the difference between empirical and conceptual statements (Cfr. Steenbergen 2004, p. 165–166).

[18] Which of course is a hint to Kant's three famous critiques.

[19] 'Kynos' is 'dog' in ancient Greek.

philosophical practitioner cherished a persistant habit of reacting physiologically instead of dialogically: "I pissed on the man who called me a dog. Why was he so surprised?"(Cited in Krajewski 1989, p. 61). Sloterdijk also uses hermeneutics as part of his kynical effort to counter-act cynicism amongst contemporary intellectuals:

> Ideology critique raises a claim that it shares with hermeneutics, namely, the claim to understand an "author" better than he understands himself. What at first sounds arrogant about this claim can be methodologically justified. Others often really do perceive things about me that escape my attention—and conversely. They possess the advantage of distance, which I can profit from only retrospectively through dialogic mirroring (Sloterdijk 1983/1987, p.19).

Important to notice are the inverted comma's flanking the word 'author'. For Sloterdijk this terms is far from unequivocal and denotes a wide range of possibilities for interpretation, from peer reviewed scientific writings and high literature down to pulp fiction and anecdotal gossip.

This contextualized reflection is also put into practice in Sloterdijk's *You must Change your Life*. There he tries to develop a kynical philosophy of sport. The title of his book is derived from the sonnet *Archaic Torso of Apollo*, which opens Rainer Maria Rilke's cycle *New Poems: The Other Part* (1908). Rilke was a roaming poetic soul with a slender physique who episodically suffered from a melancholic disposition (*spleen*). During 1905 and 1906 he assisted the famous French sculptor Auguste Rodin as a private secretary. Not knowing how to find poetical inspiration again, Rilke seeks advice from his admired athletic and virile temporary employer, whose famous, contemplative statue *The Thinker* is at the same time a paragon of muscularity, a human portrait of an 'explosive' nature. Rodin advises Rilke to step out of his depressed and passive mode, to go out, look around and simply produce, write, re-write and so gradually improve his literary skills. Rodin especially advises him to visit museums and zoos, in order to observe concrete *things.* Go out and work. *'Travailler, toujours travailler!'*

Rilke accepts the challenge, visits the Louvre, and becomes inspired by an impressive torso that is known as the *Torso of Milete*. The fact that the highly damaged stone clog probably does not represent Apollo, as Rilke thought, makes no difference, since in ancient Greece Gods were athletes and vice versa. Though head, limbs and genitals are largely absent, the athletic and muscular trunk of stone, blessed with a six-pack, sends out a call which inspires the initially dumbfounded poet to one of his most famous 'thing-sonnets'(Zwart 2010, p. 73).

> Archaic Torso of Apollo
>
> We never knew his head and all the light
> that ripened in his fabled eyes. But
> his torso still glows like a gas lamp dimmed
> in which his gaze, lit long ago,
>
> holds fast and shines. Otherwise the surge
> of the breast could not blind you, nor a smile
> run through the slight twist of the loins
> towards that centre where procreation thrived.

Otherwise this stone would stand deformed and curt
under the shoulders' transparent plunge
and not glisten just like wild beasts' fur

and not burst forth from all its contours
like a star: for there is no place
that does not see you. You must change your life
(Cited in Sloterdijk 2009/2013, p. 21)

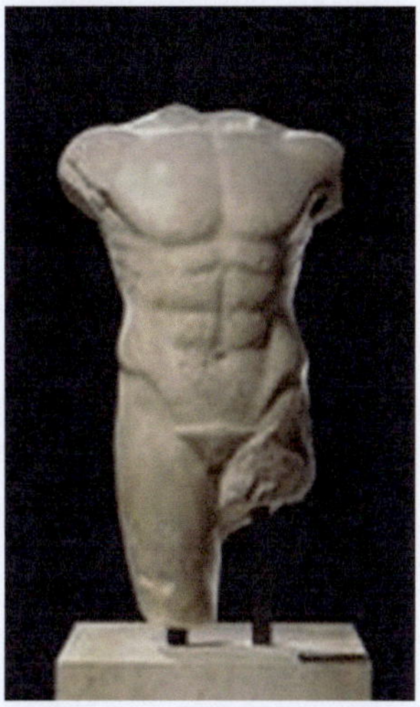

An impressive and inspiring torso at the Louvre

Sloterdijk transforms this literary anecdote into a philosophical perspective on the *conditio humanis* at large. Under the heading *Command from the Stone* he argues that Rilke's point is not that the statue is a petrified echo only interesting for those who are initiated in the world of ancient Greece, 'the humanistically educated'. It is rather a thing-construct that is suddenly 'in the air' again, to phrase it in metabletical terms. What it reminds us of is that it takes quite some effort to become a full blown member of the 'planet of the practicing'. Referring to Nietzsche's *Thus Spoke Zarathustra* (2005, p. 60) Sloterdijk states that: "Practice is defined here as any operation that provides or improves the actor's qualification for the next performance of the same operation, whether it is declared as a practice or not"(2009/2013, p. 4). Therefore we should learn to revalue training as such, and not just as a means to certain ends, such as fitness, health, appearance, prestige or financial gain. The viril clog of stone sends out an appeal to all of us: Go out and simply work! *Travailler, toujours travailler!* To amplify Rodin's advice to Rilke, Sloterdijk also once more invokes Nietzsche:

Works, first and foremost!
That is to say, doing, doing, doing!
The 'faith' that goes with it will soon put in an appearance!
-you can be sure of that!
(Friedrich Nietzsche, Morgenröthe/Daybreak,
cited in Sloterdijk 2013, p. vii).

In order to learn to (re-)value the inherent quasi-Sisyphean nature of being human, we have to be patient, attentive, and persistent, and, indeed, endorse William James's 'strenuous mood'. To achieve the desired state of body and mind, Sloterdijk argues, it takes a lot of investing in *askēsis*. Although, as said, the term originally means 'practice' or 'training' in the purely technical sense, other, derivative meanings seem to prevail in the current usage of the term ascetic. It particularly refers to a simple, secluded, repetitive and rule-obeying monastic life, far from the madding crowd and close to verticality, closer to 'God'. For Sloterdijk, asceticism however represents a wide range of historical manifestations of striving for excellence through dedicated training practices. Our 'planet of the practicing' is inhabited by individuals who are constantly and relentlessly training themselves, from ancient Stoics and Cynics and mediaval monks up to modern 'Apollonian' sport professionals.

This general human ascetic predisposition may seem self-focused, but it may well have a broader scope. We train ourselves to become better citizens, contributing to a just and sustainable society. We must change our lives and strive for a new horizon of universal co-operative training practices, resulting in a 'general ascetology'.

> In order to uncover the matrix of Old and New European techniques for shaping humans, one must first examine the training centres scattered across the whole continent in which those who practice with Christ prepare for their highest *agones,* brought into shape by their abbots, pastors, saints and learned mentors. And those who have been called 'professors' since the sixteenth century were initially no more than trainers at schools of transfiguration, and those later termed 'students' were first of all seekers in whom the eros of impossibility was at work *more academico.* They yielded willingly to the illusion that it was indispensable for all advanced civilizations: that the inimitable is imitable, and the incomparable repeatable (Sloterdijk 2009/2013, p. 322–323).

These preferred training regimens, which thus differ per era, may lead to a personal *metanoia,* a turning point in life, a radical reform of habits, a total personal mitigation. On a collective level the metanoetic imperative coming from the statue may even result in a 'renaissance', which in Sloterdijk's exuberant philosophical grammar is not bound to fourteenth century Italy, but marks a broader variety of massive societal-ideological turnovers.

Roughly since Pierre de Coubertin's introduction of the modern Olympics in 1896, the athletic ideal—so predominant in classic Greek culture, drenched with *agon* and *askēsis* (Tuncel 2013)—is in the lead again. "[A] transformation best described as a re-somatisation or a de-spiritualisation of asceticisms"(p. 27). The analogy between forms of sport and forms of discourse and knowledge should therefore be taken as literally as possible. This re-introduction of the powerful body however doesn't mean that the rational and sensitive soul is put aside. *Physis* and *psyche* are two sides of the same coin for the modern athlete.

> The completion of the renaissance through the return of the athlete around 1900 encompasses the return of the wise man: in the *panathlon* of intelligence, he makes his own contribution to clarifying the form in which that renaissance continues today (Sloterdijk 2009/2013, p. 155).

This rebirth of the muscular body also gives room to vent our '*Vertikalspannung*' (vertical tension), our inescapable but (in our horizontal society) often suppressed tendency to always try to better ourselves, as exemplified in the motto of the modern Olympics *Citius, Altius, Fortius!*

> By producing new configurations between contemplation and fitness, the current 'renaissance' enables new festivals on the plateau of the mountain of improbabilities. Anyone who has ever taken part in such a festival knows that neither a 'knowledge society' nor an 'information society' exists, as much as the new mystifiers might speak of them. What has constantly been arising since the renaissance is a multi-disciplinary and multi-virtuosic world with expanding limits of ability (p. 155).

The quintessential point of this inexhaustible human potential for pushing back frontiers is that we first have to become fully aware of our ascetic disposition as such.

> For the primal ethical imperative 'You must change your life!' to be followed, therefore, it is initially necessary for the practising to become aware of their exercises as exercises, that is to say as forms of life that engage the practising person. The reason for this is self-evident: if the players are themselves inescapably affected by what they play and how they play it (and how it has been drilled into them to play it), they will only have access to the bridge of their self-change by recognizing the games in which they are entangled for what they are (p. 145).

The tension between training programmes as such and their implicit transformative power leads Sloterdijk to refute Wittgenstein's statement that trying to define sport and games amounts to nothing but tautological language games. Wittgenstein may be right when he argues that the meaning of a word is determined by its actual usage, but what he according to Sloterdijk seems to overlook is that the decisive factor is the refinement of that usage. Once internalised, asceticism implies freedom and a high potential for change. "This supposed end in itself is, in truth, the medium in which the conversion of possessed rule-applications to free exercises take place"(p. 145).

In sum, for Sloterdijk ascetic practices are, though mediators, never sheer means to straightforward hedonistic ends, such as buzz-words often used with regard to contemporary massive sport activities (e.g. 'interaction', 'communication', 'health', etc.) suggest. Or even worse: the 'enhancement industry'with its "departments of plastic surgery, fitness management, wellness service and systemic doping" (p. 338). Well-performed and well-understood practices must also touch upon something essential that we seem to have lost on our way to *homo compensator* in technotopia.[20] Namely, understanding that life is not a one-dimensional tailor-made training

[20] Sloterdijk is anything but a dystopian or cynical thinker when it comes to technology, as already argued in the introductory Chap. 1, by referring to his concept of human-friendly homeo-technology. The term 'anthropotechnics' in the subtitle *You Must Change Your Life* also refers to this optimistic basic attitude toward technological progress. "Salvation, for Sloterdijk, lies in just the area where Heidegger believed perdition lay: that is, in the realm of technology. Yet technology, for Sloterdijk, seldom has to do with machines. It is mental and spiritual technology that interests him: the techniques with which human beings have historically made themselves secure on the Earth" (Kirsch 2015).

programme that linea recta leads to personal growth, but rather a quasi-Sisyphean effort which at a certain point may offer a glimpse of unconcealed being. This is why people are still inclined to invest so much energy in sportive practices with so little net result.

> All somewhat advanced civilizations make use of the observation that every active person is dyed in the lye of their activities until the miracle of 'second nature' takes place and they perform the near-impossible almost effortlessly.
>
> The highest theorem of explicit training theories, then, is that ability subjected to persistent furthering tension produces, almost 'of its own accord', heightened ability (p. 321).[21]

5.4 Upwardly Oriented Hermeneutics

You must Change your Life takes up the old adage 'know thyself', but combines it with the still echoing call from the stone ahtlete: work on thyself. As argued, this blueprint of a general ascetology paves the way for a synthetic re-interpretation of sport as something which is both a narrow *autotelic*, self-referential activity and a *heterotelic*, knowledge-seeking pursuit. While most philosophers of sport focus on exemplary cases in elite sport, Sloterdijk's asceticism denotes a much wider perspective on humans as natural born trainees in times of an athletic renaissance. From cripple and hunger-artist to Johan Cruyff, Lionel Messi, Eddy Merckx and Chris Froome, and everyone in between: we are all ascetics who strive for our personal best by slowly pushing the limits towards the seemingly impossible.

One may argue, nonetheless, that Sloterdijk's 'kynical' dialogical hermeneutical and often eclectic philosophy renders his general ascetology almost immune for counter-evidence. This would result, then, in a susceptibility for the 'hermeneutical circle', i.e. 'the idea that we always understand or interpret out of some presuppositions' (Grondin 2015, p. 1). Classical hermeneutics attempted to avoid this circle by sacrificing weaker interpretations to stronger ones.[22] Yet, the twentieth century the idea that we are always prejudiced burgeoned among thinkers such as Heidegger, Gadamer and Ricoeur.

> [A]s finite historical beings, we understand *because* we are guided by anticipations, expectations and questions. For them, the key is not to escape the hermeneutical circle but following Heidegger's famous phrase, to enter into it in the right way. ... Heidegger's life-long destruction of the history of Western thought in the hope of unfolding a more original understanding of Being can indeed be seen as his way of entering into the hermeneutical circle of the understanding of Being (p. 2)

[21] A similar pneumatic insight already has been put forward by William James's by means of his on the benefits of experiencing a 'second wind' (1907, p. 7). I will come back to the Sloterdijk-James connection in the next chapter.

[22] This resembles Carl 'Watergate' Bernstein's striving for 'the best attainable version of the truth'.

In the same vein Sloterdijk refers to the concept of the *circulus virtuosos*: a complex chain of events which reinforce themselves through a feedback loop with favourable results.[23] By recognizing the upwardly oriented spiral nature of the virtuous version of the circle "it becomes explicable how accomplishment leads to higher accomplishment and success to expanded success" (Sloterdijk 2009/2013, p. 321). This transformative power of well-understood asceticisms may, as argued, result in a *metanoia* on the personal and a renaissance on the collective level. But first of all we have to master ourselves and change our life. "I am to develop into a fakir of coexistence with everyone and everything, and reduce my footprint in the environment to the trail of a feather"(p. 449). The next step is to recognize the need for a "horizon of universal co-operative asceticisms" and "take on the good habits of shared survival in daily exercises"(p.452).

In an interview given on the occasion of the Dutch translation of *You must Change your Life* Sloterdijk concludes that he indeed sees a renaissance coming up of a young and dynamic avant-garde that propagates an ecological and 'active' life-style: less meat and cars and more veggies and bicycles. He even suggests that the immense popular gyms can serve as a giant 'muscular reactor'[24] (Giesen 2011). Let us apply the compelling petrified plea for a robust life-style change to a specific case: endurance sport, notably cycling, which seems highly repetitive and monomaniac, and thus not very 'playful'.[25] However, it has a relatively low impact on the limbs, and while it does not require highly developed motor skills,[26] this modern asceticism is within reach of the masses.[27] Sloterdijk reasons that especially the vertically challenged climber who, ahead of the pack, leaves all mortals far behind, deserves our respect as the contemporary substitute of the ancient god-athlete.

Sloterdijk admits "that those parts of the sporting world closest to the 'circus' in the ancient sense"—in particular the Olympics and professional football and cycling—are prone to a "result fetishism that absolutely rivals the compulsive product-oriented thinking of the economic sphere". Still he also remarks: "… what does this mean if, on the other hand, statistics show that in those sports there are ten thousand amateurs or more for every professional?" (Sloterdijk 2009/2013, p. 213).

---

[23] As opposed to the *vicious* circle, which has negative results, since it unavoidably leads to an infinite regress.

[24] A charming, but naïve thought, which has to be refuted for reasons of energetic inefficiency.

[25] Except for the childhood phase of learning how to ride a bike through trial and error. Every child remembers the excitement of its first wobbly ride without side wheels. It goes, it goes!

[26] In the seventies the Dutch cyclist Joop Zoetemelk wasn't able to absolve a very elementary obstacle-run during a TV-show, nor was he able to pull himself up just once at the horizontal bar. Still he won the Tour de France in 1980.

[27] And actually is. According to the KNWU (Royal Dutch Cycling Union) in the Netherlands approximately 10% of the population can be qualified as a sports-cyclist, while on average every Dutch(wo)men cycles 878 kilometres per year, commuting, holiday-cycling etc. included.

Ascending Colle de Nivolet (Italian Alps, 2612 m) with local amateurs. (Own collection)

In his phenomenological reflections on cycling Nepper Larsen advances the idea of the cycling subject as an "inter-being that faces an ocean of interpretations"(2010, p. 29). In the long run the eudaimonistic and environmental benefits of a cycling life cannot be justified by sheer utilitarianism. He justifies cycling in a deeper, 'ecosophical' manner by pointing out that, in contrast to a modern sports utility vehicle, a bicycle is the perfect 'anthropotechnical' (or homeo-technical!) extension piece of the human body. Once you have experienced this possibility of overcoming yourself by pushing the pedals, there is no way back. Larsen bends Descartes' *cogito ergo sum* in an ascetological direction.

> 'I pedal therefore I am.' Many things can be questioned, but one thing is beyond any doubt: in the beginning there was real motion and not just the symbolic motion of my thoughts a I carry out sedentary office work. Unimpededness is our primary state (p. 28).

Dedicated amateur Nepper Larsen, initially undaunted, but in the end frozen to the marrow after climbing Puig Major on atmospherically unstable Mallorca in April, agrees with Sloterdijk, who contends that mountains are not to be messed with: either you climb them or you leave them alone. Real ascetics are prone to go upwards, not just onwards. "Emanating from the mountains of the world, a vertical imperative hits the horizontal human. Pull yourself together, leave the lowlands and conquer the summit!"(p. 30).

Sloterdijk's practical and practicable ascetology evidently provides an escape route for the dull horizontality of every day life in an ecosophical way. Those who learn to agonize themselves in order to optimize their ascetic potentialities, will in the end become better human beings—step by step. His dialogical hermeneutics shows us that virtuous, upwardly oriented spiral cycling is a preferential form of asceticism that can be performed in the great wide and preferably hilly open at your own pace. By cultivating the good habit of cycling as a relatively sustainable 'anthropotechnical-homeotechnical' means, we not only learn to fight the struggle with and against ourselves. As upwardly tending endurance athletes we also learn to transform our often uncompromising attitude towards our natural environment into a coexistent ecosophical agony with the world.

We are all natural ascetics born into a world filled with un-metaphorical hindrances. In a meaningful ascetic life, cycling is an involuntary attempt to overcome necessary obstacles.

References

Caddick, N., and E. Ryall. 2012. The social construction of 'mental toughness' – A fascistoid ideology? *Journal of the Philosophy of Sport* 39 (1): 137–154. https://doi.org/10.1080/00948705.2012.675068.

Edgar, A. 2013. A hermeneutics of sport. *Sport, Ethics and Philosophy* 7 (1): 140–167.

Fry, J. 2014. Sport. In *The Bloomsbury companion to the philosophy of sport*, ed. C. Torres, 371–373. London: Bloomsbury.

Fusche Moe, V. 2014. The philosophy of sport and continental philosophy. In *The Bloomsbury companion to the philosophy of sport*, ed. C. Torres, 52–65. London: Bloomsbury.

Giesen, P. 2011. Peter Sloterdijk, Duitse filosoof: Naar de sportschool om de planeet te redden. *De Volkskrant*, April 30 2011: 7–8.

Grondin, J. 2015. What is the hermeneutical circle? (First draft of an essay to be published) In: *The Blackwell companion to hermeneutics,* Keane N. and C. Lawn eds., (2016). Available at https://www.academia.edu/13944229/What_is_the_hermeneutical_circle?auto=download&campaign=weekly_digest. Accessed 3Sept 2015.

Heidegger, M. 1927/2008. *Being and time*. New York: Harper & Row.

Huizinga, J. 1955. *Homo ludens – A study of the play-element in culture*. Boston: Beacon Press.

James, W. 1907. The energies of man. *The American magazine* (reprint). Available at: https://archive.org/stream/energiesofmen00jameuoft/energiesofmen00jameuoft_djvu.txt

Kirsch, A. 2015. *Against cynicism – A philosopher's brilliant reasons for living*. Available at: http://www.newrepublic.com/article/113387/peter-sloterdijks-philosophy-gives-reasons-living

Krajeweski, B. 1989. Critique of cynical reason, Peter Sloterdijk. *Journal of the Midwest Modern Language Association* 22 (1): 61–64.

Loland, S. 1995. Coubertin's ideology of Olympism from the perspective of the history of ideas. *Olympika: Nternational Journal of Olympic Studies* (4): 49–77.

———. 2003. *Technology in sport: Three ideal-typical views and their implications*. Available at: http://idrottsforum.org/articles/loland/loland_2.html

Lopez Frias, J. 2014. Internalism. In *Bloomsbury companion to the philosophy of sport*, ed. C. Torres, 362–364. London: Bloomsbury.

Lopez Frias, J., and X. Montfort. 2015. A hermeneutical analysis of the internalist approach in the philosophy of sport. *Physical Culture and Sport* LXVII: 5–12.

Lunt, D., and M. Dyreson. 2014. A history of philosophic ideas about sport. In *The Bloomsbury companion to the philosophy of sport*, ed. C. Torres, 17–37. New York: Bloomsbury.

McNamee, M. 2015. *The spirit of sport and the World Anti-Doping Code. Abingdon*. Routledge.

Meier, K. 1988. Triad trickery: Playing with sport and games. *Journal of the Philosophy of Sport* 15: 11–30.

Nepper Larsen, S. 2010. Becoming a cyclist – Phenomenological reflections on cycling. In *Cycling philosophy for everyone: A philosophical tour de force*, ed. J. Ilundáin-Agurruza and M. Austin, 27–39. Chichester: Wiley Blackwell.

Nietzsche, F.W. 2005. *Thus spoke Zarathustra: A book for everyone and nobody*. Oxford/New York: Oxford University Press.

Piacente, A. 2015. Reverse play: Towards a philosophy *from* sport. *Sport, Ethics and Philosophy* 9 (1): 58–75.

Reid, H.L. 2009. Sport, philosophy, and the quest for knowledge. *Journal of the Philosophy of Sport* 36 (1): 40–49.

Sloterdijk, P. 1983/1987. *Critique of cynical reason*. Minneapolis: University of Minnesota Press.
———. 2009/2013. *You must change your life – On anthropotechnics*. Cambridge/Malden: Polity Press.
Steenbergen, J. 2004. *Grenzen aan de sport. Een theoretische analyse van het sportbegrip*. Maarssen: Elsevier.
Suits, B. 1978/2005. *The grasshopper: Games, life, and utopia*. Ontario: Broadview Press.
———. 1988. Tricky triad: Games, play, and sport. *Journal of the Philosophy of Sport* 15: 1–9.
Tuncel, Y. 2013. *Agon in Nietzsche*. Milwaukee: Marquette University Press.
Wittgenstein, L. 1953/1978. *Philosophical investigations*. London: Blackwell.
Zwart, H. 2010. Wordt ons leven anders? Over genomics en zelfarbeid – Beschouwingen naar aanleiding van Peter Sloterdijks *Du musst dein Leben ändern. Tijdschrift voor Gezondheidszorg en Ethiek* 2 (20): 73–76.

Chapter 6
Continental Pragmatism: Enduring Life in the Strenuous Mood

The higher heroisms and the old rare flavors are passing out of life

(William James 1899, p. 6).

Abstract Philosophy is often epitomized as the noble art of asking the right questions. In this chapter I will also try to formulate a sport philosophical answer to the question how we are to live in times of environmental crisis and moral desorientation. I will do so by broadening the practical philosophical perspective I developed so far. Firmly rooted in continental philosophy, over time I have increasingly become infected by William James's pragmatist adage that truth can only be found in the practical consequences of philosophical thinking. Integrating the pragmatic stance into my continental approach, I now will argue in favour of a life fully lived in strenuous endurance sport, for I regard both traditions as complementary rather than exclusionary. Endurance sport, conceived as a committed and holistic lifestyle, rather than as a gratuitous playful pastime, is a preferential tool for carving out the good life we are to lead, and which leads into a sustainable future.

As will have become clear in previous chapters, for my continentally inspired view on the benefits of human endurance in general I am indebted both to Sigmund Loland's ecosophical work and to Peter Sloterdijk's kynical thoughts on how to change our lives through asceticism. In *You must Change your Life*, Sloterdijk regularly hints at possible directions for improving our lives. We should become more environmentally conscious, less susceptible for the temptations of hyper-consumptive modern life—"banalised Enlightenment", in Sloterdijk's wording —, willing to put in more effort when it comes to satisfying our vertical needs, becoming more resilient, mentally as well as physically. Still, there remain quite a few loose ends when it comes to concretely stepping over from theoretical ascetology to ascetic action. The tenacious question still is: how to concretize and materialise ecosophical-ascetological initiatives?

R. Welters, *Towards a Sustainable Philosophy of Endurance Sport*, Library of Ethics and Applied Philosophy 37,
https://doi.org/10.1007/978-3-030-05294-2_6

My aim in this chapter is to further develop Sloterdijk's provocative, agonistic style in a pragmatist manner: how to change our lives for the better through properly practiced endurance sport, particularly of the cycling kind?

Keywords Continental philosophy · Pragmatism · Strenuous mood · Endurance

6.1 For the Better

In the introduction of *You must Change your Life* (2009/2013) Sloterdijk unfolds the moral bottom line for his proclamation of an 'anthropotechnic turn'. Since God officially has been declared dead in 1882 by Friedrich Nietzsche in *The Gay Science*, divine salvation is no longer a prospect. As humans, we are ourselves in charge for changing the world for the better, he contends.

> [T]he tiresome distinction between 'true religion' and superstition loses its meaning. There are only regimens that are more and less capable and worthy of propagation. The false dichotomy of believers and unbelievers becomes obsolete and is replaced by the distinction between te practising and the untrained, or those who train different (Sloterdijk 2009/2013, p. 3).

The resulting conception of 'anthropotechnics' roughly equates all kinds of refined ascetological practices that have been developed over time to optimize life, ranging from sheep-herding and biotechnology to physical training. Anthropotechnics, then, is a homeo-technological means that attempts to follow the curved lines of nature to the flourishing ecosophical ends explored by Arne Naess and Sigmund Loland.

Sloterdijk's eloquently amplified plea is an emergency call that similes the notion of 'meliorism'. According to Jonathon Kahn (2009), this striving for betterment refers to "society's innate, inexorable tendency towards improvement"(p. 37). Yet, the idea that things tend to get better on their own accord must be considered a so-called 'naturalistic fallacy'. It is fallacious to explain that which is good reductively in terms of natural properties (Moore 1903). The 'good' is not a natural phenomenon that simply wells up, improvement requires active investment on our part. The crucial point of upwardly tending asceticism is precisely that we have to put in serious effort ourselves to make things better and more ecosophical. As already argued: in the Anthropocene we are responsible ourselves. Thus, once more: *Travailler, toujours travailler!* Works, first and foremost!

Also Kahn is reluctant to consider meliorism a 'natural' phenomenon that simply blossoms without proper care. He therefore endorses William James's pragmatic variant of meliorism, in which the idea of human hope is put forward as a more obvious enhancer for the desirable societal change. For James the idea that things tend to get better by itself equals non-committal 'optimism'. True pragmatic meliorism requires a substantial effort. Referring to James's *The Varieties of Religious Experience* Kahn says:

> Pragmatic meliorism is hopeful in girding itself for battle with real factors and forces that make the universe a seriuos place of real loss: 'He is willing to live on a scheme of uncertified possibilities which he trusts; *willing to pay with his own person, if need be*, for the realization of the ideals which he frames.' As James says, pragmatic meliorism 'treats salvation as neither inevitable nor impossible. It treats it as possibility, which becomes more and more of a probability the more numerous the actual conditions of salvation become.' The hope of these ideals is born of strife and striving (2009, p. 38).

While James's pragmatic variant still echoes a sense of religion, Sloterdijk's plea for drastic change is rather a self-initiated strategy of self-improvement.

Sloterdijk's anthropotechnic metanoetic asceticism has a stronger connection with pragmatism than the sparse explicit references to James in *You must Change your Life* suggest, however. Following the pragmatist take on endurance sport developed by Douglas Hochstetler and Peter M. Hopsicker (2012), I argue that, in order to understand the full meaning of endurance sport, we, indeed, have to become true runners and real cyclists, rather than dabbling joggers and occasional weekend warriors. The authors argue that there can, or perhaps even should, be seasons of life when athletes engage in endurance sport in a monomaniac manner, to prepare for a marathon for instance, or any other long-distance endurance event. They warn, however, for "the potential dark side of this pursuit of excellence" (2016, p. 335). This means: becoming obsessed by training and racing, and, as a consequence, relinquishing other, social interests in life. To carve out "an Aristotelian Golden Mean of sorts between minions and puny fellows[1]" (p. 336) they argue for a proper balance between a monomaniac dedication to endurance sport on the one hand and taking care of relationships and social responsibilities on the other.

In this chapter I will develop a more individualistic, even somewhat anti-social take on endurance sport as a preferential tool for strenuously materializing the ideal of human flourishing. Decidedly developing ones very own personal 'vertical challenge' is the incentive for the more durable lifestyle we need on the collective level. We need to push ourselves uncompromisingly to our very limits for a better world. Sloterdijk's specific brand of continentalism and a lived-through variant of pragmatism will go hand in hand when it comes to understanding a sporting life fully lived in endurance. A practical and practicable philosophy that provides some idea of an answer to how we are to live for the better, this is the heart of the strenuous matter.

6.2 Finding the Right Gear

Before starting to pedal for durable meaning in a strenuous mood, a preliminary note on the following subjective narrative seems appropriate. Sympathizing with Heidegger's previously mentioned famous remark in *Being and time* on the indeterminateness of *Das Man*— "Everyone is the other, and no one is himself" (1927/2008,

[1] "Yonder puny fellow, however, whom every one can beat, suffers no chagrin about it, for he has long ago abandoned the attempt to 'carry that line', as the merchants say, of self at all" (James 1890/2007, p. 330).

p. 165)—[2] I now will follow his call for a more subjective, 'earthy', and 'caring' mode of philosophy. This results in a well-informed philosophical outline for a desirable common future for all creatures, great ánd small. It is the effort we are willing to put in that determines the feasibility of the beckoning perspective. Therefore I will stress the merits of the participative stance in the field of sport philosophy, especially in the field of cycling in the mountains, the ultimate fight against back-pulling forces, which, I contend, probably only can be understood in its full agonistic depth when experienced personally. Eventually, however, as already noticed in Chap. 2: there is always the duty to return from the anecdotal to the general ecosophical ascetological stakes of the current study.

I will now begin with some anecdotal observations *ad hominem*.[3] In a revelatory interview concerning the *Tour de France*, Sloterdijk, who elaborately argues for living a life in appropriate physical practice, reveals himself as a practitioner of the strenuous kind (Gorris and Kurbjuweit 2008). As mentioned, being a runner in his younger days, he later changed to cycling, which for him represents the modern variant of ancient heroism. Especially the vertically challenged climber who, ahead of the pack, leaves all flatly living mortals far behind, deserves our respect as a contemporary substitute of the ancient god-athlete. Together with a Dutch friend[4] Sloterdijk in 2007 actually managed to climb the feared and loathed Mont Ventoux, a famous winding and windy ascent in the Provence of 21 km with an average grade of 7.5%, in about 2 h and a half, which is to be considered as quite good for a man of his stature and age. When the interviewers ask him if this ascent gave him a sense of what a professional cyclist goes through, he confides the following:

> More than a sense: You realize that the feat these men perform is completely beyond the grasp of ordinary mortals. It's almost like studying theology. You need to attain the first degree of initiation to understand that you don't understand anything (Gorris and Kurbjuweit 2008).

[2] "We take pleasure and enjoy ourselves as *they* [man] take pleasure; we read, see, and judge about literature and art as *they* see and judge; likewise we shrink back from the 'great mass' as they shrink back; we find 'shocking' what they find shocking. The 'they', which is nothing definite, and which all are, though not as the sum, prescribes the kind of Being of everydayness" (Heidegger 1927/2008, p. 164).

[3] In the next chap. I will also provide anecdotal evidence *ad feminem*.

[4] René Gude, the late Dutch 'philosopher of the fatherland', who, in the line of reasoning of *You must Change your Life*, considered philosophy "as a training programme for life" (Van de Poll, 2015) Cfr. also Sloterdijk's diaries for his cycling experiences (2018).

Anonymous cyclists at the lunar landscape of Mont Ventoux, the bald mountain, the hell of the south

This sense of feeling nothing but an insignificant dabbler, a novice in the world of real suffering, is corroborated by an amateur cyclist who was interviewed right after climbing Mont Ventoux.

> "Gradually my speed drops to 6mp and everything starts to close around me. All I can see is faded whiteness, and even after 3 km the weather station appears no nearer. I round another bend which fails to flatten as I had hoped and it crossed my mind that I can't do this for much longer—I can't even spare the breath to take a badly needed drink (diary, 12 July 2003)" (Spinney 2006, p. 709).

This kind of humble reflection makes Sloterdijk's outline for a general ascetology imaginable and practicable for mortals. It is as if he wants to say: asceticism is within reach for all of us, look at me! While the majority of sport philosophers still concentrate on highly commercialised elite sport, Sloterdijk in *You must Change your Life* also concentrates on the lower echelons; the modest variants that enable us to become our very own personal Hector, Achilles or Penthesilea. When it comes to true asceticism, well-willing, benevolent amateurism has to be preferred over elitarian result fetishism.

Also Heither Reid, already mentioned as considering sport a potential truth seeker rather than an innocent pastime, does not shy away from weaving personal cycling experiences into her philosophical discourse. She clarifies her comparison of the classic qualitative Olympic ethos—qualified as *poionistic* in the metabletical Chap. 4—and the current *posonistic* "efficiency ethos" (Reid 2017, p. 159) in sport, which these days also flourishes among 'amateur' aficionado's, by referring to a personal anecdote:

> Having climbed the spectacular Gavia pass[5] in the Italian Alps this past summer, I compared notes at dinner with a fellow cyclist. What I remembered was the beauty of the mountains and having overcome the doubt about whether I could make it to the top. His own experience was a product of the power and heart rate monitors attached to his bike—he talked about watts, gear ratios and heart rate data, never mentioning the environment or even his state of mind. … Quantitative epistemology tends to set aside, disregard, and sometimes even deny the existence of that which cannot be quantified—and so much of what is valuable in sport is qualitative rather than quantitative. A healthy sports ethos will take note of the quantitative, but will not let it squelch out the qualitative value of the sport experience (p. 164).

During his unnerving ascent of the Mont Ventoux as a *theologia negativa* Sloterdijk homepathically experienced what it must be like to be a pro. Reid, who, as said, nearly qualified for the 1984 and 1988 Olympics as a track cyclist, and who still runs a company for assisted cycling tours in Italy, probably has had ascetic experiences which brought her much closer to the Gods of yesteryear.

This difference in physical aggregation state doesn't matter that much for the ascetological-ecosophical stakes pursued in the current study, however. It is the very idea of the willingness to put in a substantial physical effort to overcome vertical challenges that makes the striving for a durable change for the better worthwhile, regardless of whether we do this as a well-willing dabbler or as a proto-professional. A personal empiric turn makes the broader idea of sport philosophy truly practible, beyond the often vented ifs and buts of the generalised I,—the loathed *moi commun* or *Das Man*. Instead, the concrete physical *me* must be put on a pedestal again.

Standing on the shoulders of the before-mentioned philosophical giants makes me less hesitant to provide a short note on my own practical ascetological-ecosophical gear and the accompanying practices. I own and frequently use seven bicycles, ranging from a rickety city-bike and a so-called 'world-traveller' to a mountainbike and a high-end and meticulously maintained time-trial machine.[6] I also own some seven pairs of running shoes. Recently I also bought a pair of goggles and swimming trunks, so I can, once I have mastered the art swimming,[7] finally step over from long distance cycling and running to the real thing: the ironman distance triathlon. Preferably a mountainous one.

---

[5] The Passo de Gavia is the tenth highest pave road in the Alps (Height: 2621 m. Length: 17.3 k. Average gradient: 7.9.). The pass is often on the route of the Giro d'Italia. Sometimes the pass is designated the *Cima Coppi*, the highest point of the race. One of the most heroic stages in the Giro took place at 5 June 1988. The race passed over the Gavia in a snowstorm, which resulted in an epic stage won by Dutchman Erik Breukink. American Andrew Hampsten, the second-place finisher, became the overall race leader and finally won the Giro. Many cyclists had to step out of race at the top of the Gavia due to nearly frozen limbs. Others, such as the Dutchman Johan van der Velde, who started the climb in a leading position, were transported in cars towards the finish, turned a blind eye by the officials.

[6] I do not own a drivers license, however, which ecosophically levels out the vast arsenal of cycles.

[7] Which is no sinecure as of a certain age because of declining motor ability (if any).

As for my own competitive level and aspirations: during races my first objective is always simply to finish. But usually I'm also striving to improve my personal best, or even secretly hoping for a podium place in my age-group. Most of the time I am simply training, however. Sometimes in a small group, but preferably on my own, in the woods and on quiet and narrow roads, paved and unpaved, engaged in Rousseau-like conversations with myself as a solitary runner. Enduring hardship, but usually enjoying it. Trying to find the right mean between *poson* and *poion*, between investing in the right gear and putting in loads of training effort and enjoying the environment and a heightened state of mind. Nearly every day combining preaching with practicing, that is what makes my ascetic life worthwhile.[8]

As for my sportive ecosophical aspirations: as already explained in Chap. 3, Loland's ecosophical questions always resonate somewhere in the back of my head. This results in the three following mantra's:

1. *Whatever your result in a race, every time you push your limits there is sustainable growth.*
2. *Determine your own right mean between investing in high end technology and training hours.*
3. *Understand that it takes a considerable effort to become a true and dedicated endurance athlete.*

And once again, especially the bicycle has specific, well-rounded benefits to offer when it comes to practice ecosophy. This extemely efficient high-tech artefact is the real existing alternative that enables almost all people[9] to get around, to travel, to go up and to reach for the divine in a sustainable manner. The bicycle is, shortly, a perfect tool in an increasingly polluted and deadlocked world. So, hop on your bike, chose the right gear, and climb your own mountains. Or, as the abbess phrases it in the *Sound of Music*:

> Climb every mountain,
> Search high and low,
> Follow every byway,
> Every path you know.
> Climb every mountain,
> Ford every stream,
> Follow every rainbow,
> 'Till you find your dream (Metrolyrics 2013).

---

[8] I'm in good company. The Belgian philosopher Marc Van den Bossche suggests that the university might donate him a Colnago racing bike, since this the perfect tool for him to develop interesting ideas on. He does so with a twinkle in the eye and in a popular publication. But still. "Science will thrive by this!" (2005, p. 133, my translation).

[9] In the Netherlands there is a vast market for so-called special needs bicycles for people with specific handicaps, such as amputated limbs or balance disorders.

Climbing the Galibier (2,556 m, French Alps) during a cycling vacation from The Netherlands to the South of France. (Private collection)

6.3 Reconsidering the Watershed

As already touched upon in the previous chapter, the supposed distinction between analytic and continental academic philosophy is "very limited if not inaccurate"(Fusche Moe 2014, p. 53), since both originate from the same European philosophical tradition, in particular Immanuel Kant's transcendental philosophy. Fusche Moe furthermore stipulates that many philosophers living in mainland Europe are dedicated to the analytic cause, while quite a few North American philosophers openly converted to continental philosophy. Still, analytic philosophers share "the problem-oriented and scientific-empirical outlook of the natural sciences" while continental philosophers are more prone to understanding "what appears to be meaningful for a person in the sense that it has tried to comprehend life as it is lived from the first person point of view" (p. 53).

Still recognizing the blurring boundaries between analytically inclined narrow internalism, or formalism, and 'continental' broad internalism, touched upon in the previous chapter, I claim that the philosophy of sport is still fairly preoccupied with the Suitsian idea of sport as the voluntary and playful attempt to overcome unnecessary obstacles, with a (often hidden) preference for the (alleged autotelic) amateur-stance. This un-committal and innocent playful 'grasshopperian' attitude seems not very appropriate for the case at hand, though: striving for a better world by means of endurance sport, with a special focus on hard (but not per se elite) cycling. Although moments, or even longer periods of lightness, flow and pleasure may occur, these sports that take a deep breath are, above all, a matter of monotonous repetition and perseverance rather than joyful, unconcerned play. These stamina sports are about enduring life in the long run with an occasional glimpse of unveiled truth, rather than experiencing sheer joy. I argue that Sloterdijk's suggestive ascetological findings are of utmost importance to fully assess the deeper meaning of endurance sport.

The question that now muscles in is: how to anchor the striving for sustainability (by means of a properly understood and ditto practiced personalised asceticism) in a pragmatically reinforced continental, broad internal framework? How to intertwine Sloterdijk's hermeneutic exuberance with James's practical engagement and direct style?

Although it is difficult to place the various versions of philosophical pragmatism under one heading, it differs from the precise and truth-seeking analytic tradition by arguing that the meaning of a proposition is determined by its use in linguistic practice and that there is no such thing as absolute truth. In this sense pragmatism concurs with the relativist undertones of hermeneutics, which argues that there may be quality differences in interpretation and levels of understanding, but there is no final truth.

However, while Sloterdijk's approach bears certain family resemblances with both hermeneutics and pragmatism, there is also a beckoning *telos* in his exuberant witticisms, namely the mentioned striving for betterment in the Anthropocene era. Philosophy has to become an active force again in the struggle for ecosophy. This activism is an important thread in Sloterdijk's oeuvre. In his *Critique of Cynical Reason* (1983/1987), his European breakthrough as a public philosopher, he already sighed that contemporary philosophy is in deep agony.

> Faced with its demise, it would like now to be honest and reveal its last secret. It confesses: The great themes, they were evasions and half-truths. Those futile, beautiful, soaring flights – God, Universe, Theory, Praxis, Subject, Object, Body, Spirit, Meaning, Nothingness – all that is nothing. They are nouns for young people, outsiders, clerics, sociologists (p. xxvi).

More specifically Sloterdijk criticizes contemporary societal cynicism. According to him this is a defeatist strategy which inoculates itself against critical resistance, and thus must be considered a case of "enlightened false consciousness"(p. 19). As explained in the previous chapter, to counter this widespread negativism, Sloterdijk calls for a more authentic, thriving and 'doggish' attitude, as exemplified by the lifestyle of the ancient 'kynic' Diogenes of Synope, the down-to-earth philosopher who lived in a water barrel on the Athens market square, asked Alexander the Great to step out of his way because he obscured the daylight, masturbated in public, and allegedly pissed on the man who called him a dog. Under the heading *In Search of Lost Cheeckiness* Sloterdijk reveals the constructive undercurrent of a plain but philosophical kynical life lived in happy refusal, which is: enlightenment through unmasking. This results in a down-to-earth post-transcendental philosophical style.

> Ethical living may be good, but naturalness is good too. That is all kynical scandal says… Here begins a laughter containing philosophical truth, which we must call to mind again if only because today everything is bent on making us forget how to laugh (p. 106).

In his much-debated essay *Rules for the Human Zoo: A Response to the Letter on Humanism*[10] (1999/2009) Sloterdijk seems to have become somewhat defeatist, however, if not cynical in the loathed modern sense. He now argues that since the humanizing effect of wise books appears to have expired, we should consider more deliberate means to ensure the preservation of Aristotle's characterization of man as a *zoon politikon,* a political animal.

Using the word 'taming', and suggesting that the upcoming new genetic technologies deserve a more upright discussion of 'bio– cultural' reproduction, caused a huge controversy in Germany, still struggling with its Nazi-history. Especially Jürgen Habermas (1929–), the grand old man of German post-war leftism, publicly accused Sloterdijk of re-introducing the idea of eugenics.[11]Although Sloterdijk in *Rules for the Human Zoo* certainly plays around with uncomfortable ideas and controversial issues, in particular eugenics, I would argue that his open ended essay also can be interpreted as an ultimate, daring, albeit somewhat desperate attempt to revitalise the humanizing power of the written word. Two thousand years after Plato, according to Sloterdijk, wisdom has become obsolete.

---

[10]This refers to Martin Heidegger's *Letter on Humanism* (1947/1977). There Heidegger argues that classic humanism has failed since it has itself in the centre of being by interpreting *homo sapiens* as an *animal rationale*, trying to control reality instead of listening to depth being in a passive mood (*Gelassenheit* or releasement).

[11]Cfr. for a detailed analysis of this debate Couture (2016), who in Chap. 4 *Controversy* refers to Sloterdijk's militant nature: "he wishes rather to embody an agonistic stance, inhibited by neither political sensitivity nor scholarly caution, that dares to publicly raise controversial issues". This suits perfectly well with the overarching idea of this study: to stand up for a revaluation of a physically oriented philosophy of sport that can serve as a grindstone for a sustainable lifestyle beyond the narrowing perspective of Habermas's drawing table theory of communicative action, which largely ignores the idea that physicality is prior to deliberation. Sloterdijk is far more aware of the whims of our limbic system. Shortly after the first skirmishes as a result of *Rules for the human zoo* journalist Frits de Lange has called Sloterdijk "an easy rider philosopher, born to be wild" in the Dutch national newspaper *Trouw* (October 9, 1999). It has to be noted, however, that Sloterdijk is a rebel with a cause. To channel this Sloterdijkian intellectual wildness into a durable lifestyle that still leave space for our inborn competitiveness is another way to phrase the key issue of this book.

> What is left to us in the place of the wise is their writings, in their glinting brilliance and their increasing obscurity. ... Letters that are not mailed cease to be missives for possible friends: they turn into archived things. ... Perhaps it occasionally happens that in such researches in the dead cellars of culture the long–ignored texts begin to glimmer, as if distant light flickers over them. Can the archives also come into the Clearing?[12] Everything suggests that archivists have become the successors of the humanists. For the few who still peer around in those archives, the realization is dawning that our lives are the confused answer to questions we were asked in places we have forgotten (1999/2009, p. 27).

6.4 Into the Pragmatic Mood

In *You must Change your Life* Sloterdijk has recovered his positive activism again, although he regularly is fobbing the reader off when it comes to concrete action. Classic pragmatism, with its predilection for functionalism and consequentialism, may help to concretize Sloterdijk's 'metanoetical' and 'renaissancistic' message of taking on the good habits of shared survival in daily exercises. William James (1842–1910), one of the founding fathers of philosophical pragmatism, contends that there is no transcendental 'truth' but nothing but a 'reality' that efficiently has to be dealt with. He also coined the idea of 'truth in consequences'. This refers to the conviction that when different approaches lead to the same desirable result, both are true.

It has to be noted, however, that James' probably best-known phrases "truth's cash value" (1907a, p. 200) and "the true is only the expedient in our way of thinking" (p. 222) often are used out of context. Their meaning is far more subtle than bluntly stating that any idea with some practical utility is 'true'. This is more 'crude pragmatism': the view that the right approach "is the one that 'gets results', most obviously on the scoreboard or on the clock" (Devine and Knight 2017, p. 35).

James's subtle pragmatic ponderings rather resemble Sloterdijk's process-oriented ideas on asceticism. James stresses the importance of the *process* from beliefs—or "thoughts in rest"(1901/1902, p. 293)—to true actions. (In our case this implies: switching over from Rilke's passive melancholic mood to Rodin's active, virile athleticism.) Sloterdijk briefly refers to James's idea that conversions may also happen in a "sick soul" or a "divided self", even "without any religious turn" (Sloterdijk 2009/2013, p. 306). He furthermore pays credit to the nineteenth century pragmatist conviction that the self-production of humans "is made explicit via study of the *vita activa*"(p. 320). And, finally, he also mentions the "piecemeal supranaturalism" which "perfectly suited the pragmatic immanentism of the Modern Age" (p. 371), both put forward in James's *The Varieties of Religious Experience* (1901/1902).

The relation between Sloterdijk's post-transcendental hermeneutic continentalism (with a touch of moralism) and James's pedagogic interpretive pragmatism is fairly intricate, however. First, one should note the striking similarity between

---

[12] An open space in the woods, according to Heidegger the place where truth can be revealed (*aletheia*).

Sloterdijk's idea that training hard results in increased resilience and heightened ability and James's ponderings in *The Energies of Man* on the benefits of a second wind (or 'second breath' as the Dutch saying goes), which befalls us when we push ourselves to (and over) our limits.

> But if an unusual necessity forces us to press onward, a surprising thing occurs. The fatigue gets worse up to a certain critical point, when gradually or suddenly it passes away, and we are fresher than before. We have evidently tapped a level of new energy, masked until then by the fatigue– obstacle usually obeyed. There may be layer after layer of this experience. A third and a fourth 'wind' may supervene (James 1907b, p. 7).

Also, James's idea of conversion (either religious or profane) implies the possibility for relieving one's vertical tension through striving for a 'personal best'. Evidently humans are prone to save their own skin in the vortex of being. But at the same time our arduous training practices may cherish a sense of modesty. Furthermore James acknowledges that a personal optimum often is the best we can get. "[W]hen we touch our own upper limit and live in our own highest center of energy, we may call ourselves saved, no matter how much higher someone else's centre may be" (James 1901/1902, p. 165). Sloterdijk tends to strip conversions from their religious connotations. Changing your life is a matter of getting on your feet and practicing like hell.

> The whole complex known as ethics comes from the gesture of conversion to ability. Conversion is not the transition from one belief system to another; the original conversion takes place as an exit from the passivist mode of existence in coincidence with the entrance into the activist mode. It is the nature of the matter that this activation and the avowal of the practising life comes to the same thing (Sloterdijk 2009/2013, p. 195).

When we dialogically and consequentially reconsider the given that today humans are natural born athletic ascetics with a vertical tension that somehow has to be vented, we cannot but conclude that seriously practiced endurance sport is an *in*voluntary attempt (to turn Suits's before-mentioned phrase of game-playing as the voluntary attempt to overcome unnecessary obstacles) to overcome a necessary obstacle towards a meaningful and enduring life.

Nevertheless, the academic philosophy of sport usually still concentrates on ludity. In *Agon in Nietzsche* Yunus Tuncel reminds us of the fierce roots of competitive physicality, however: "With the ancient Greeks, the agonal spirit had already existed in poetry, mythology, arts, and *athletics* (italics mine), before it re-produced itself in thought with the rise of philosophy" (Tuncel 2013, p. 255). In the 1880s, some 20 years before Rilke became re-inspired by an archaic torso with an eye-catching six-pack, Nietzsche feverishly attempted to resurrect this ancient spirit in order to unmask the feebleness of his own era. Over a century later Tuncel sighs that "agonal feelings do not have a sufficient, wider acceptance today, sometimes not even among the competing forces" (p. 257). Nietzsche's ideal of a strong agonistic individualism has been reduced to the playful "demands of the populace for self-preservation, for mundane affairs, for sameness and equality; monolithicity rules…. What counts is the excitement of the masses, a petty spectacle of scoring goals" (p. 257).

By performing agonal contests on the razor's edge, sport, nevertheless, still can be a powerful means to a serious end: seeking unpolished, immediate knowledge, cultivating virtue and enhancing the good life. This comes down to fully opting for the challenging life William James argued for after experiencing too much numbing complacency during a conference organised in a peaceful scholarly retreat in upstate New York:

> At Chautauqua there were no racks, even in the place's historical museum; and no sweat, except possibly the gentle moisture on the brow of some lecturer, or on the sides of some player in the ball-field.... It looks indeed, thought I, as if the romantic idealists with their pessimism about our civilization were, after all, quite right. An irremediable flatness is coming over the world.... The higher heroisms and the old rare flavors are passing out of life (James 1899, p. 6).

Real life needs vertical challenges and needs to be confronted with the fresh but sometimes scouring wind of society. This is also the case when it comes to finding a proper balance between institutionalised definitions of health—say, exercise moderately for 150 min per week[13]—and pushing the limits of supposed physical frontiers, with the risk of overtraining, injury and social isolation.

6.5 Into the Strenuous Mood

In *The Heights of Humanity: Endurance Sport and the Strenuous Mood* (2012) Douglas Hochstetler and Peter Hopsicker pick up the plea for high spirits and a matching physique by arguing that endurance sports have special benefits to offer when it comes to overcoming the horizontal dimension of contemporary life. For them there is a fundamental difference between occasional joggers and riders ('breadth') and dedicated runners and cyclists ('depth'). They argue that the classic philosophical pragmatism of William James and John Dewey is the key to understanding why these sports can make life significant and worthwhile. Strenuously practiced endurance sport opens possibilities for 'recovering humanity' in a manner qualitatively different from the 'transcendental approach' towards sport as distracting from everyday dullness (p. 118). This specific brand of 'transcendentalism' Hochstetler and Hopsicker refer to has been developed in the later 1820s and 1830s in the United States. Building on Immanuel Kant's concept of 'transcendentals' as intuitive but necessary conditions for knowledge, representatives such as Ralph Waldo Emerson and Henry David Thoreau rely on (transcendental) subjective intuition rather than on the earthly empiricism of the pragmatists.

---

[13]This is the official norm set by the Dutch government. This norm is specified for different age-groups. For 55 up the advice is to exercise minimal 5 but preferably 7 days a week moderately intensive, which means walking at 3–4 km per hour or cycle at 10 km per hour (Gezondheidsraad 2017). (Which is quite slow, in a flat country as The Netherlands. Even with a reasonable headwind it should be possible to ride 15km per hour, I argue).

Hochstetler and Hopsicker agree with Anderson's (2001) transcendentalist line of thought that anyone who leisurely and occasionally participates in sport is susceptible to the possibility of growth through commitment. They contend, however, that only those who are truly dedicated to the endurance case have a higher potential for personal growth. "Because of their engagement, these athletes have a *heightened* chance of meeting possibility, a *heightened* chance of establishing creativity, and a *heightened* chance of learning about self" (Hochstetler and Hopsicker 2012, p. 120). To substantiate their claims they work up the idea of a process-oriented 'increased ownership' for serious practitioners who consider sport more than an appropriate tool for temporary "distancing" (Kretchmar 1982) themselves from the daily drag. True dedication and serious suffering are preferential tools for creating a meaningful life by trying to improve one's personal best.

> Both the jogger and runner may find their respective movement outlets enticing as a way to combat the restlessness of routine life. However, the runner even has more reason to anticipate movement and endurance sport, not only as a respite from the work world but also in the recognition that through these efforts one gradually develops an identity as an endurance athlete (Hochstetler and Hopsicker 2012, p. 121).

Other than ball games, sprinting and sports that require highly developed motor skills, endurance sports also have special benefits to offer for the ageing athlete: "many runners and cyclists can achieve their personal best times during their 30s and even their 40s" (p. 121).

There is a downside to dedication and patience, carved in the sporting subject over decades of training, however. Whereas youth normally comes with agility, joy and freedom from worry, age implies less flexible limbs, and more need of physical care, practical wisdom, moderation, punctuality, meticulous devotion and abstinence of all sorts, and therefore the danger of living the life of a monomaniac. "It is certainly true that sport can become mechanistic, and on occasion a setting where individual agency becomes squelched" (p. 122).

The pitfall of narrowing down life to sheer asceticism however doesn't mean that endurance sport is dull routine *per se*. There is still an openness for creating a certain distance to the daily drag in the strenuous mood. A life almost fully lived in training programs still "has the potential to produce deeply meaningful experiences for those athletes called to sporting practices", Hochtsteler and Hopsicker argue (2012, p. 122). Those who are able to endure life the hard way have a unique chance to "drive life into a corner" and "season life", they argue."While these qualities are available to the weekend warrior, we contend that physical activity, conducted in the strenuous vein, is a fertile ground for experiencing the depth of our humanity" (p. 132).

Also Ilundaín-Agurruza points at the specific advantages of a lived through pragmatic general ascetology to further develop a full-fledged philosophy of (endurance) sport. He concludes that James's argument for the strenuous life of sometimes "reads like an analysis of athletic forays".

> [H]e speaks of asceticism in terms of an organic hardiness … that shuns too much ease, a dietary temperance and non-pampering of the body, one that beckons a sacrifice for a god highlights the obsession with a challenge to overcome…. He pointedly notes how this leads to 'exercises' prompted by genuine perversions where pain is felt as pleasure …. If we exchange the 'higher being' for the notion of a 'calling,' we have a list of prerequisites to truly engage in the aforementioned pursuits and a way to explain for that perversion that Nike's 'No Pain, No Gain' so pithily embodies. The intrinsic enjoyment that many of us *habit*ually find on the slopes, waters, turfs, courts, roads, gymnasiums, dojos, dance studios, and workshops is predicated on nurturing a toughness that battens the hatches against the lure of sedate, painless, and oftentimes mind-numbing hedonistic leisure outlets the likes of which TV consumption or endless hours of online 'activity' illustrate. An ascetic disposition—not at odds with enjoyment of a refined kind—lies at the heart of successful habit. And this happens in and through engaged and living bodies (2014, p. 260).

6.6 Potential Perils Rebutted

The decisive point now is: how far can we go, when it comes to living life in tough and persistent practices? What are the limits—if any—of the ascetic disposition?

In *Normative Concerns for Endurance Athletes* (2016) Hochstetler and Hopsicker tone down somewhat their previous pragmatist plea for the strenuous mood. Now they argue in favour of a more holistically oriented attempt to fulfill Aristotle's notion of *eudaimonia*, or human flourishing, which is the final goal in his virtue ethics. Other than gasping joggers, somewhat corpulent weekend warriors and easy-going breaststroke swimmers, long distance runners, serious cyclists and smooth crawlers may face a 'potential dark side' of their monomaniac training programs. By becoming glassy slaves of the repetitive rhythm of absolving daily mileages and personalised training programs, the truly dedicated ones are prone to neglect their social environment. Frenetic fanatics may "relinquish other interests in pursuit of excellence in their particular sport" (p. 335), such as the potential perils of damaged interpersonal relations and marital problems. "To avoid becoming either a minion or puny fellow, one must find some way to navigate the various tensions inherent with endurance sport participation" (p. 347).

In the same eudaimonistically inspired way, Gunnar Breivik (2010) argues for putting the all too strenuous sporting mood into perspective. He proposes well-roundedness and concentration as alternatives for the strive for excellence: "perfection is an attractive choice but no obligation"(p. 87). The idea of a well-balanced Golden Mean—say between recklessness and cowardice—is prominent in Aristotle's virtue ethics. Especially the notion of *phronēsis*, or practical wisdom, is appropriate for the case at hand: how much strenuousness can we endure in our search for the good life? It should be reminded, however, that, similar to Sloterdijk and James, Aristotle in his *Nicomachean Ethics* already has argued for an excellence-oriented agonistic approach in sport:

> And as in the Olympic Games it is not the most beautiful and strongest that are crowned but those who compete (for it is some of these that are victorious), so those who act win, and rightly win, the noble and good things in life (Aristotle 1999, p. 13).

Trying to win, getting the best out of yourself, pushing yourself to your limits is not just a matter of narcissism, it also refers to the intrinsic value of the strenuous life. Besides the danger of social isolation, Hochstetler and Hopsicker warn of the possibility for endurance athletes to develop the well-known over-training syndrome (2016, p. 346). Also Sloterdijk acknowledges the perils of living the strenuous life close to, or sometimes even over the edge. Still, if well-performed, the positive effects of super-compensation after extreme training sessions prevail.

> The rhythms of regeneration hold the secret of the overexertion that leads to higher performance levels. This phenomenon has been intuitively comprehensible since time immemorial, and had already been exploited for intensive training session in antiquity; on the other hand, the ancients were also familiar with the phenomena of overtraining that appear if regeneration rhythms had been disregarded (Sloterdijk 2009/2013, pp. 320–321).

Of course, overtraining has to be avoided under all circumstances, but tentatively finding out about your own limits makes an ascetic life worthwhile. Micro-cracks in muscles that are put to a severe test cause soreness. This soreness is a a sign to temporary slow down. Train hard, rest hard. The strenuous life is about learning to know thyself.

The distinctive advantage of endurance sports as a preferential mould for venting our ascetic tendencies is that these are within reach of the masses. Elite runners, cyclists and tri-athletes of course have special talents that are beyond the reach of normal mortals. But those who are willing to spend sufficient training hours and develop stamina may also enter the Hall of Fame of long distance race finishers, or perhaps even become age-group winners. Here again the overarching adage is: try hard to get the best out of your very self, rather than overtaking opponents or winning a specific contest. Triathlon-trainer Tim Heming (2015) argues that:

> [A]ll of us, whatever level we are, aim to race faster, higher and stronger…The breadth and depth of triathlon experience can be truly exceptional. ... Don’t get caught up in others’ targets. Instead, create your own triathlon journey and remember to enjoy it (pp. 78–79).

Of course one may discern an occasional smile on a competitor’s face during a long distance endurance event. My educated guess is, however, that endurance aficionados enjoy their sport in a far more modest manner than, say, soccer or basketball players, with their vehement outbursts in the petty spectacle of scoring goals. The stealthy smile of an endurance athlete is an accidental belch drenched in a sea of indefatigable resilience. “Satisfaction, that’s the word. Not just pleasure” (Elcombe and Tracey 2010, p. 256).

Not just pleasure. (Source: Roadcycling)

Precisely this orientation towards acquired stamina rather than innate talent makes endurance sport easier to perform than for instance soccer or gymnastics. Especially cycling has special benefits to offer for the crowd. The noble art of pedalling exerts a relatively low impact on the limbs, and does not require highly developed motor skills, which usually are necessary for playing games at a serious level. In the seventies the Dutch cyclist Joop Zoetemelk wasn't even able to perform a rather elementary obstacle-run during a TV-show, nor was he able to pull himself up at the horizontal bar. Still he won the Tour de France in 1980. Meanwhile also professional sport has undergone a thorough change of training methods. Also core-stability, functional strength and flexibility are important issues, these days. But the major asset still is: pushing the pedals as hard as possible.

On top of it, widely practiced cycling has more than just a sportive potency. It might help us to realise the massive turn towards a more sustainable lifestyle we are so desperately in need of. Hopping on a bike more often certainly would make the world a better place. Creating your very own upwardly oriented challenge, following the commandment from the stone to elevate yourself, definitely has more potential for depth than unconcernedly pedaling or jogging along. Really pushing yourself to your limits will quite probably indeed result in Hochstetler's and Hopsicker's idea of heightened chances of learning about the self through tough stages of "increased ownership". Even more, once infected with the endurance-virus, even somewhat puny fellows may become dedicated amateurs of the better sort.

When it comes to actual improvement of our lives in Sloterdijk's re-birthing vein, we should keep in mind however that competitive sport, even in its mildest manifestations, to a certain degree is also always about competitiveness, which according to Loland (1996, 2001, 2006) may be considered the very metaphor for our unsustainable lifestyle. But as Sloterdijk emphasizes: for every perfidious professional there are a thousand well-willing amateurs. Properly performed endurance sport nevertheless can help to overcome the tension between competiveness and coexistence. By revaluing the concept of *askēsis* as a contemporary and relatively clean and green means to fight the struggle with and against oneself, we may overcome the uncompromising fight against our natural environment and transform the human condition into a coexistent agony wíth the world, as already proposed in Chaps. 2 (*Sport and the Environment: Considering Sustainable Thoughts*) and 3 (*Answering Three Ecosophical Questions: Asceticism*).

Nevertheless, we have to avoid using endurance sport as a straightforward means to an in itself praiseworthy end. There should always remain a nucleus of indeterminateness in our congenital tendency for asceticism, beyond the idea of linear growth towards a better and fuller self, and in the long run perhaps even a better world. True asceticism is not about using proper means to straightforward ends, Sloterdijk argues. It is rather about learning to endure life as such:

> But if man genuinely produces man, it is precisely not work and its concrete results, not even the 'work on oneself' so widely praised in recent times, let alone through the alternatively invoked phenomena of 'interaction' or 'communication': it is through life in forms of practice (2009/2013, p. 4).

In a similar way James (1907b) underlines the mentioned beneficial experience of a second wind, which also helps to improve every next performance of the same operation, so that the "higher heroisms and the old rare flavors" (p. 6) just will not pass out of life. To counter to the flatness that indisputably is coming over the world, we should put ourselves on the rack, not just every now and then, but on a basis as regular as possible.

The net sum of the preceding encounter of continentalism and pragmatism is that we are all natural born ascetics in a world filled with hindrances. Since they learn to endure life with all its defects, those who learn to really agonize themselves in order to optimize their innate ascetic potentialities will become better humans in the end. Strenuously performed sports that take long breath—competitive or not—are a preferential tool for carving out some idea of enduring meaning. This easily accessible, but at the same time quasi-Sisyphean form of asceticism, should be performed in great wide and preferably mountainous open spaces. Only there we can escape the daily grind of the horizontal life. Rather on your own or in small group, at your own pace, but if possible fairly ahead of the pack and the peloton.

By revaluing the quasi-Sisyphean concept of *askēsis* as a contemporary and relatively clean green means to fight the struggle with and against oneself, we may overcome the uncompromising fight against our natural environment and transform the human condition into a 'coexistent agony' wíth the world. This, however, without throwing the competitive, proto-agonistic baby that still lingers in us out with the bathwater.

> We do not necessarily see the connection between competitive practices and transfiguration[14] of cruelty,[15] violence and suffering. Perhaps agonism is not a sufficient response to all human suffering; however, it is one response. Just like the tragic human, it accepts human suffering at least in two areas; it channels cruelty and destruction unto culturally accepted arenas and it enables humans to deal with loss and death. It must also be mentioned that agonism promotes strong individuality, the best response to human suffering (Tuncel 2013, pp. 256–257).

All things considered, only a physically strenuous life is worth to be lived. To turn Bernard Suits's gratuitous definition of game-playing around once more: in a meaningful life, endurance sport is an involuntary attempt to overcome necessary obstacles. Transposed in cycling terms this results in the idea that the emptiness in the faces of non-racers is shocking, and not the other way around. Or such as Albert Camus ends his essay *The Myth of Sisyphus* (1942/1991):

> I leave Sisyphus at the foot of the mountain! One always finds one's burden again. But Sisyphus teaches the higher fidelity that negates the gods and raises rocks. He too concludes that all is well. This universe henceforth without a master seems to him neither sterile nor futile. Each atom of that stone, each mineral flake of that night filled mountain, in itself forms a world. The struggle itself toward the heights itself is enough to fill a man's heart. One must imagine Sisyphus happy (p. 24).

The lonely climber on the steepest of slopes must be considered as being on his or her way to a meaningful and flourishing life.

---

[14] Sloterdijk argues that until the rise of European universities education was transfiguration-centred. "And those who have been called 'professors' since the sixteenth century were initially no more than trainers at schools of transfiguration, and those later termed 'students' were first of all seekers in whom the eros of impossibility was at work *more academico*. They yielded willingly to the illusion that it was indispensable for all advanced civilizations: that the inimitable is imitable, and the incomparable repeatable" (Sloterdijk 2009/2013, p. 322–323).

[15] This refers to Nietzsche's idea of *'Verfeinerte Grausamkeit'*, refined cruelty, as a positive, character-building sentiment that is worth striving for.

Steep section during a trail run at Lanzarote: a struggle filling a man's heart. (Source: 3athlon)

References

Anderson, D. 2001. Recovering humanity: Movement, sport and nature. *Journal of the Philosophy of Sport* 28: 140–150.

Aristotle. 1999. *Ethica Nicomacheia*. Kitchener: Batoche Books.

Breivik, G. 2010. Philosophical perfectionism – Consequences and implications for sport. *Sport, Ethics and Philosophy* 4 (1): 87–105.

Camus, A. 1942/1991. *The myth of Sisyphus and other essays*. New York: Vintage Books.

Couture, J.P. 2016. *Sloterdijk*. Cambridge: Polity Press.

Devine, J.W., and C.J. Knight. 2017. Beyond 'crude pragmatism' in sports coaching: Insights from C.S. Peirce, William James, and John Dewey: A commentary. *International Journal of Sports Science & Coaching, 12* (1): 35–37.

Elcombe, T., and J. Tracey. 2010. Strechted elastics, the Tour de France, and a meaningful life. In *Cycling philosophy for everyone: A philosophical tour de force*, ed. J. Ilundáin-Agurruza and M. Austin, 241–252. Chichester: Wiley Blackwell.

Fusche Moe, V. 2014. The philosophy of sport and continental philosophy. In *The Bloomsbury companion to the philosophy of sport*, ed. C. Torres, 52–65. London: Bloomsbury.

Gezondheidsraad. 2017. *Beweegrichtlijnen 2017*. https://www.gezondheidsraad.nl/nl/taak-werkwijze/werkterrein/preventie/beweegrichtlijnen-2017. Accessed 23 Feb 2018.

Gorris, L., and D. Kurbjuweit. 2008. *Philosopher Peter Sloterdijk on the Tour de France: 'The riders are just regular employees.'* Der Spiegel Online International. http://www.spiegel.de/international/europe/philosopher-peter-sloterdijk-on-the-tour-de-france-the-riders-are-just-regular-employees-a-565111.html Accessed 6 Mar 2012.

Heidegger, M. 1927/2008. *Being and time*. New York: Harper & Row.

———. 1947/1977. *Letter on humanism. Basic writings*. London: Routledge.

Heming, T. 2015. Join the club. *220 Triathlon* March 2015: 78–79.

Hochstetler, D., and P. Hopsicker. 2012. The heights of humanity: Endurance sport and the strenuous mood. *Journal of the Philosophy of Sport* 39 (1): 117–136.

———. 2016. Normative concerns for endurance athletes. *Journal of the Philosophy of Sport* 43 (3): 335–349.

Ilundáin-Agurruza, J. 2014. William James—Pragmatic pioneer. *Sport, Ethics and Philosophy* 8 (3): 258–270. https://doi.org/10.1080/17511321.2014.981359.

James, W. 1890/2007. *The principles of psychology*, vol. I. New York: Cosimo Classics.

———.1899. *What makes a life significant?* http://philosophy.lander.edu/intro/articles/jameslife–a.pdf.

———. 1901/1902. *The varieties of religious experience. A study in human nature*. https://worldu.edu/library/william_james_var.pdf. Accessed 5 Feb 2018.

———. 1907a. *Pragmatism, an new name for some old ways of thinking. Popular lectures on philosophy*. New York: Longmans, Green, & Co.

———. 1907b. The energies of man. *The American magazine* (reprint). https://archive.org/stream/energiesofmen00jameuoft/energiesofmen00jameuoft_djvu.txt. Accessed 5 Feb 2018.

Kahn, J. 2009. *Divine discontent: The religious imagination of W.E.B. Du Bois*. Oxford: Oxford University Press.

Kretchmar, S.R. 1982. Distancing': An essay on abstract thinking in sport performances. *Journal of the Philosophy of Sport* 9: 6–18.

Loland, S. 1996. Outline of an ecosophy of sport. *Journal of the Philosophy of Sport* 23: 70–90.

———. 2001. Record sports: An ecological critique and a reconstruction. *Journal of the Philosophy of Sport* 28: 127–139.

———. 2006. Olympic sport and the ideal of sustainable development. *Journal of the Philosophy of Sport* 33: 144–156.

Metrolyrics. 2013. *The sound of music lyrics.* http://www.metrolyrics.com/climb-every-mountain-lyrics-the-sound-of-music.html. Accessed 15 Apr 2018.

Moore, G.E. 1903. *Principia ethica.* http://fair-use.org/g-e-moore/principia-ethica. Accessed 16 Feb 2018.

Reid, H.L. 2017. Why Olympia matters for modern sport. *Journal of the Philosophy of Sport* 44 (2): 159–173.

Sloterdijk, P. 1983/1987. *Critique of cynical reason*. Minneapolis: University of Minnesota Press.

———. 1999/2009. Rules for the human zoo. A response to the *Letter on humanism. Environment and Planning D: Society and Space* 27: 12–28. https://rekveld.home.xs4all.nl/tech/Sloterdijk_RulesForTheHumanZoo.pdf. Accessed 5 Feb 2018.

———. 2009/2013. *You must change your life – On anthropotechnics*. Cambridge/Malden: Polity Press.

Spinney, J. 2006. A place of sense: A kinaesthetic ethnography of cyclists on Mont Ventoux. *Environment and Planning D: Society and Space* 24: 709–732.

Tuncel, Y. 2013. *Agon in Nietzsche*. Milwaukee: Marquette University Press.

Van den Bossche, M. 2005. *De passie van Marc Van den Bossche: Wielrennen*. Rotterdam: Lemniscaat.

Van de Poll, W. 2015. René Gude (1957–2015): "Ik ben echt niet bang voor de dood". *Trouw* 14 March. https://www.trouw.nl/home/rene-gude-1957-2015-ik-ben-echt-niet-bang-voor-de-dood-~a2f1a2e5/. Accessed 17 Feb 2018.

Chapter 7
On *Agon* and Ecosophical Endurance: Finding Your Own Pace

Pain is just a detail.

(Former professional cyclist Gert-Jan Theunisse in Creteur 2010, p. 124)

Abstract Prompted by Sigmund Loland's ecosophy of sport and Peter Sloterdijk's analysis of human beings as upward tending training animals, as well as by insights from historical phenomenology (or 'metabletics'), hermeneutics, interpretivism and pragmatism, I have argued for a vertically challenged life, preferably to be played out on two unmotorized wheels. Following this plea for cultivating stamina, I ended the previous chapter with an ode to the strenuous mood of the lonely endurance athlete, ahead of the pack, tired but satisfied, and potentially on the brink of the good and flourishing life.

Now it is time to return to the overall objective of this study. How can endurance sport at large, and cycling in particular, despite their challenging nature, contribute not only to self-knowledge, but also to self-improvement and sustainability? How to step over from individual *metanoia* to a collectively (re)introduction of good ecosophical habits? This chapter provides a closer analysis of endurance sport as a preferential contemporary manifestation of *askēsis*, or training. Once again, but now empirically more fortified, I argue that practiced at your own pace and understood as a process of gradual self-improvement, rather than winning over opponents per se, endurance sport can overcome the stalemate between exclusionary competitive sport and striving for a more inclusionary sustainability. Experiencing and overcoming agony will turn out to be a necessity for the radical change of lifestyle we need.

I will concretise and materialise my ecosophical-ascetological stakes and claims by zooming in on specific cases of high suffering in endurance sport. This practical philosophical descent in the pain-cave will result in an illuminating vista on the wider ascetological-ecosophical implications of long distance endurance sport. Agonistic sport and environmental sensitivity, the twain shall meet and merge into a consequential truth that takes a serious effort.

Keywords Agony · Self-improvement · Endurance · Consequential truth

R. Welters, *Towards a Sustainable Philosophy of Endurance Sport*,
Library of Ethics and Applied Philosophy 37,
https://doi.org/10.1007/978-3-030-05294-2_7

7.1 Sport in Agony

In *The Universal Dictionary of the English Language* 'agony' is defined as "a struggle for victory in the games; anguish of mind … Extreme bodily pain or physical pain or suffering … The last death struggle, the pangs of death" (Wyld 1958, p. 21). In his study *Agon in Nietzsche* (2013) Yunus Tuncel notices that in ancient Greece, 'agon' refers to "the event of the gathering" as well as "the actual acts of competition, the games, the festival, or the spectacle itself" (p. 11). Later, agon came to refer to the ancient Greek culture of competitiveness, ranging from "the dialogue in tragedies to debates within the court setting"(p. 11). Clarysse and Remijsen argue that, since the whole Greek culture was permeated with a competitive spirit, the ancient Greek predilection for sport contests is not accidental. Agonism was an integrated part of education. The young were encouraged to compete, to become the best. This ideal of overcoming mediocrity was already prominent in the *Iliad* of Homeros, the manual book for Greek moral education. Through physical training, the young were taught to become resilient and never to give up. Losing simply wasn't an option. This did not imply egocentricism, however, the researchers argue. This 'verticalism' also could benifit the Greek community as a whole.

> That competition was a central aspect of Greek culture and of the life of every Greek also found its expression in the Greek language. Greek has one word covering all forms of competition: *agon*. This could mean 'war', but also 'dispute', philosophical, political or juridical, or 'contest'. There were not only contests in sports, but also in music and drama, between potters and even between doctors. And in every contest participants tried to be the best and, in this way, to achieve honour. The Greek word for this ambition or urge to compete, *philotimia*, can be translated literally as 'love for honour' (Clarysse and Remijsen).

This agonistic furor in ancient Greece seems absent in most contemporary sport practices. Nonetheless, I will argue that the auto-agonistic tendency in contemporary long distance endurance sport not only sheds light on what this type of sport is about, but also can be a test-case for an inclusive and updated reading of 'agony'.

A renewed plea for a spirit of competitiveness will make many people frown, especially in the green camp. In Chap. 2, I already indicated that because of its inherent competitiveness, sport usually has a negative connotation in environmental philosophy or 'ecosophy'. From a sustainability perspective, sport coincides with the relentless and hyper-consumptive way we deal with 'nature' during the Anthropocene. Taken as such, elite sport indeed seems to equate the ethos of excess: always striving to win, no matter what the cost. To counter this cynically amplified token of the law of diminishing returns, I argued (in Chaps. 2 and 3 and elsewhere) for a more inclusive and broader ecosophical perspective on sport by proposing the ascetic bicyclical human as the adequate contemporary measure of all things. In the resulting asceto-logical-ecosophical point of view, agony still resonates, be it somewhat toned down and domesticated. Now the time has come to return to some stronger manifestations of the agonistic undercurrent of strenuously experienced physicality.

Contemporary philosophy of sport largely seems to have dropped the idea of agon, however. In Chap. 5 (*Ascetic Practices, Hermeneutical Cycles and Ecosophical Endurance*) I argued that many sport philosophers are still preoccupied with Bernard Suits' intrinsic definition of game-playing, put forward in *The Grasshopper: Games,*

Life, and Utopia (1978/2005): "Playing a game is the voluntary attempt to overcome unnecessary obstacles" (p. 55). This definition has to be interpreted in the heuristic triangle of play, games and sport. Play, then, refers to the basic human need for distraction from working life. Games are formalized forms of play. And even competitive sports are eventually considered as playful games.

Initially Suits suggested that 'play' is the common denominator in all these activities. In *Tricky Triad: Games, Play, and Sport* (1988) he, however, refines his case, by contending that only un-committal amateur sport may be playful, whilst professional sport per se is not. Professional sport is instrumental. Moreover, in the modern Olympics[1] Suits discerns "*a kind of compulsion* to win that is absent in a friendly game of tennis, or a pick-up game of baseball or hockey"(p. 8). Following this line of reasoning, 'real sport' according to Suits has an intrinsic or 'autotelic' value; it is just played for its own sake.

By and large, Suits, and in his slipstream the greater part of other narrow internalists (Cfr. Lopez Frias 2014, p. 363), implicitly still seem to avow to the un-committal playful stance of the amateur-gentle(wo)man. All too serious sport is considered extrinsic, as a means toward other ends, or 'heterotelic', and thus contrary the rather opaque concept of 'the spirit of sport' (Cfr. McNamee 2012, 2014), the idea that there is (or at least ought to be) a silent code of desirable behaviour beyond the codified rules of a specific game. Probably this should refer to something like 'playing it hard but fair', not using doping, or the 'voluntary suspension of play' (kicking the ball out for treatment of an opponent, and receiving the ball back after the re-start), etc.

This gentle-(wo)manlike lusory attitude still reflects the sentiment that 'real sport' has to be played in a gentleman-like athmosphere; an interesting game of golf or a nice football match, with just a little bit of civil war on the pitch and a pint of beer after the specific zero sum game. This lusory attitude is exemplified by Roger Bannister, the first man who ran a mile in less than 4 min in 1954. He reasoned that athletics is a luxury and that "the essence of sports is that while you're doing it, nothing else matters, but after you stop, there is a place, generally not very important, where you would put it" (cited in Horsley and Horsley 2013, p. 33). Contemporary elite sport on many occasions has surpassed the ludic autotelic stance by far, however. It cannot be denied any longer that sport has become big business. In its utmost cynical variant, sport adheres to the following quote (incorrectly attributed to American football coach Vince Lombardi): "Winning isn't everything; it's the only thing."[2]

In the narrow internalist, or formalist, attempt (how praiseworthy in itself) to counter today's elite sport cynicism, unfortunately also the ancient need for serious competitive struggles to create moral virtue (*arēte*) has sunk in oblivion. In ancient Olympic Greece there were no friendly and un-committal ball-games. Those were the days of highly competitive and extremely serious athletic contests. Tough physical games were necessary hurdles to overcome, in order to lead a fulfilling and valuable life, rather than leisurely performed voluntary activities.

---

[1] Which were in those days for 'amateurs' only.

[2] The quote actually is attributed to UCLA Bruins football coach Henry Russell ('Red') Sanders, who in 1950 during a physical education workshop told his group: "'Men, I'll be honest. Winning isn't everything' then following a long pause, 'Men, it's the only thing!'" (Rosenbaum 1950). Vince Lombardi on several occasions used Russel's quote approvingly, however.

> The athletic contest, or *agon*, represented an important proving ground for competitors to vie for victory and its attendant *arete*. Athletic contests were cultural, social and religious events, to be taken seriously, as the term *agon*, or struggle, implies (the English word 'agony' is a derivative). The Greek verb 'to play' (*paizein*) was never used in relation to athletics. Games were not 'played'. They were struggled (Lunt and Dyreson 2014, p. 18).

This in current times seemingly suppressed serious 'classic' un-ludic take on sport is the reason why Heither Reid proposes the revitalization of sport as a "knowledge-seeking activity" (Reid 2009, p. 40) and as a carrier of social values rather than a disinterested goal in itself, with only a homeopathic hint of *agon*. And this is why, as already stipulated in the previous chapter, Yunus Tuncel (2013) reminds us of the fact that contemporary egalitarian society tends to flatten out truly lived classic Greek agony by reducing sport to bread and circuses.[3] "We have only a poor copy of agonistic justice; fairness obliterated by the modern idea of equality, which confuses non-equals and blends haphazardly the strong and the weak" (p. 257).

Drinking cup (500–475 BC) depicting the fight between two boxers and two 'pankratiasts'. Pankration, literally meaning 'all force', is a combination of wrestling and boxing. It was a dangerous sport, in which everything was permitted except biting, gouging (stabbing with your finger in your opponent's eye, nose or mouth) and attacking the genitals. On the image, the pankratiasts try to gouge out each other's eyes. The umpire is about to punish them for this offence. The objects on the wall (a bag with products for washing and anointing) situates this scene in the gymnasion. (Source: Clarysse and Remijsen, Ancient Olympics)

[3] Cfr. for a critique of the idea that sport in the boom times of the Roman empire was reduced to *panem et circenses* Heather Reid's idea that during those days in the arena the existing order was challenged as well. "Some say that we should look to Rome rather than Greece to see our own athletic values reflected in antiquity. There, they say, sports were primarily entertainment enjoyed by masses of inactive spectators and exploited by politicians who sought public favour. But even the bloody spectacle of gladiator fights preserved the truth-seeking and educational functions that connect sport and philosophy. While the Emperor saluted the Roman spectators, who were seated in tiers according to social class, the contest itself challenged that hierarchy. It gave the lowly 'socially-dead' gladiator the opportunity to prove his social worth by prevailing in a publicly observed and strictly regulated test of relevant virtues. The condemned gladiator who received the wooden sword of freedom from the emperor as the crowd shouted its approval stands as an enduring symbol of sport's ancestral ties to philosophy" (2009, p. 47).

Some may discern a germ of fascism in Tuncel's justification of Nietzsche's anti-egalitarian agonism. If we follow Sloterdijk's line of ascetological reasoning, however, the decisive point is vertical potentiality. Also hunger artists, or the armless violin player Carl Herman Unthan, who played his instrument as a virtuoso with his feet, but also tiny climbers, pared to the bone, can become the strongest in their specific niche. It is the vertical effort that sets the tone, not the pure unpolished brute talent as such. It is the will to assiduously and aggresively invest in even the slightest talent that beats the ascetological measure. It is the will to personal empowerment rather than dominating your opponents that justifies the suppressed agonal spirit.[4]

> That life can involve the need to move forwards in spite of obstacles is one of the basic experiences shared by the group of people whom, with a carefree clarity, one formerly called 'cripples', before younger and supposedly more humane, understanding and respectful spirits of the age renamed them the handicapped, those with special needs, the problem children, and finally simply 'human beings'. (Sloterdijk 2009/2013, p. 40)

Even cripples can be spurred on by this imperative, Sloterdijk kynically reasons, such as the asthenic poet Rilke, who became vertically inspired by the silent imperative emanating from the limb- and headless statue:

> In short, people had to speak about the handicapped, the differently constituted, to stumble on a phrase that expresses the general constitution of beings under vertical tension. 'You must change your life!'means, as we saw in Rilke's torso poem: you must pay attention to your inner vertical axis and judge how the pull from its upper pole affects you! It is not walking upright that makes humans human; it is rather the incipient awareness of the inner gradient that causes human sto do so (pp. 59–60).

The broader context of virtuous competitive agonistic education seems to be missing in current times indeed, such as both Tuncel and Sloterdijk suggest. There is still space and hope for befitting vertical agony, however. The concept of sport as 'knowledge-seeking' with specific benefits for experiencing physicality and embodied cognition, has regained influence in philosophical reflections on niche phenomena such as 'risk', 'adventure' or 'nature sports'. These are the hidden places where true agony and real virtue still can blossom and all horizontal values can be vertically revaluated.

Similarly the German sport philosopher and sport pedagogue Arno Müller (2004, 2008) argues that high risk sports, such as for instance 'free climbing' (mountain climbing without safety measures), reach further than 'ludic' and rule-bound zero-sum games,[5] or mass sport activities conceived as a health enhancer for the physically challenged crowd. High-risk sports may make us truly feel what it is to be *existent*; what is means 'to be towards death' (*Sein zum Tode*), in Heidegger's phrasing. Sports that explicitly seek vertical danger have a potential for becoming larger than horizontal life, Müller reasons. "I would

[4] Francis Fukuyama already suggested the re-instalment of the Platonic *thymos*, or thrift, in his *The End of History and the Last Man* (1992).

[5] One team or player wins (+1), the other loses (−1), or the game ends in a draw. The net sum of the game therefore always is 0.

like to summarize the 'gains' of these 'death-centered sports' with the term *enhanced existence* or at least *moments of enhanced existence*" (2004, pp. 60–61).

Also the Alaskan sport philosopher and outdoor sports entrepreneur Kevin Krein opens up the conception of what is usually referred to as 'sport' to wider heterotelic horizons. Nature sports seek meaning rather than victory. He argues for the prevalence of value-creation over competition in so-called 'nature sports' (mountain biking, whitewater sports, rock climbing, extreme skiing, and snowboarding, etc.): "without the imposed structure of formal competition, humans can creatively define their own values" (2014, p. 206). Also the Canadian sport philosopher Leslie Howe[6] argues that activities she calls 'remote sports' are process-oriented rather than 'agonistic' competitive strivings: "that is, about the experience itself and the development of sport-specific skills rather than the result or defeating someone else" (2008, pp. 3–4).

There is more at stake in 'risk', 'nature' and 'remote sports', however, than enhancing existence, value-creation and process-orientedness, I would argue. I see these outdoor-oriented sports are a mixture of agonistic, value creating and process-oriented motives, ranging from cutting edge competition to deep ecosophical motivations. Mountain biking, for instance, can differ from a solo activity in search of the deeper meaning of life off the beaten tracks, to an well-organized Trans Alp experience and an 'agonistic' World Cup race at the razor's edge. On the other hand, cross country mountain biking is an official Olympic discipline since Atlanta 1996, and thus 'domesticated'. But even running offers a multitude of possibilities for externalising one's inner urge to move: from record-oriented track-running and big city marathons to extreme trail-running events. On the other hand, even in 'urban sports'—according to Howe the very counterpart of remote sports, say football or basketball—one can observe 'competitors' playing the game 'process oriented' and 'just for fun' in an unofficial setting.[7]

Furthermore it has to be noted that currently many people seem to climb mountains just for plain hedonistic reasons. Take the recent rise of 'industrial' mass tourism on the Mount Everest. In 2013 there were more than 700 people that reached the summit.[8] Confronted with this sobering phenomenon, the famous Tyrolean climber

[6]Who, as mentioned in Chap. 2, does not argue that remote sport is better than conventional sport as such, but just points out that remote sports have some special benefits to offer.

[7]This doesn't have to mean that there are no rules the players of the specific 'fun'-game have to stick to, however. Dutch 'pleintjesvoetbal' (football on squares, often with lining and small goals), for instance, has a few explicit rules, but also hold strong implicit ideas on how to play the game. A cunning 'panna' (playing the ball between the opponent's legs) is sometimes considered as more valuable than scoring a simple goal. There is a shared more or less explicit aesthetic ethos of a 'spirit of sport', so to say. This makes 'pleintjesvoetbal' a combination of a zero sum game and a so-called juried sport. There is a discussion in the Netherlands on the functionality of the trick-oriented nature of 'pleintjesvoetbal' for more the more result-oriented professional football. Still, there is no question about the benefits for developing essential motor skills through 'pleintjesvoetbal', especially in current times of declining attention for physical education in primary and secondary school.

[8]On March 22, 2013 it was extremely busy on the slopes of the Mount Everest. "Yesterday climbers reported scenes of chaos as around 100 people tried to read the summit in the space of a few hours, while lengthy queues formed below trickier sections. (Venables 2013)

Reinhold Messner[9] has concluded that the term adventure tourism has become an oxymoron. “Adventure is all about risk, uncertainty and self-determination—not buying a predictable, packaged commodity: you pay me £40,000 and I’ll make your dream come true” (Venables 2013). The crossing of the boundary begins where the show stops.[10]

Now, instead of opposing highly organized competitive elite sport to broader, supposedly more process-oriented ‘natural’ conceptions of sport, I once more will attempt to overcome a binary view which hinders developing a credible and nuanced philosophy of sport. In order to overcome the stalemate between narrow autotelic and broade heterotelic perspectives, as well as between shallow green adaptation and deep ecosophical mitigation, I will broaden the contours of a durable agonistic ascetological ecosophy. This effort is undertaken for epistemological as well as normative reasons. Epistemologically, it is important because it helps us to develop an overarching definition of sport that takes us beyond the autotelic play-paradigm. Normatively, it is important because of the urgent need for a vision on sport that meets the call for sustainability in the Anthropocene.

Building on the contours of a hermeneutics of endurance sport outlined in Chap. 5 and the strenuous pragmatic reflections developed in Chap. 6, I now propose a perspectivist approach, which sees these various perspectives as complementary yet mutually challenging. This philosophical view, which goes back to Protagoras, but was rehabilitated by Friedrich Nietzsche, reasons that, because all interpretations arise from particular perspectives, there are many possible conceptual schemes, or perspectives on truth or value. Nietzsche’s version of ‘perspectivism’ counters the potential relativistic undercurrent by proposing a more committed and ‘directive’ version. Although it cannot be convincingly argued that there is one true world—the classic metaphysical stance, fiercely rejected by Nietzsche—this does not imply that all perspectives are equally valid. On the contrary, there are obviously more powerful and weaker perspectives. It is their performative power which ultimately determines the strength of specific perspectives. This post-metaphysical, post-relativist point of view is nicely captured in an aphorism in Nietzsche’s *Nachlass*, also known as *The Will to Power* (1901/1968):

> Against positivism, which halts at phenomena—‘There are only facts’—I would say: No, facts is precisely what there is not, only interpretations. We cannot establish any fact ‘in itself’: perhaps it is folly to want to do such a thing.

---

[9] Reinhold Messner (1944–) made the first (and so far the only) solo ascent of Mount Everest. He also made the first ascent of the highest peak of the world without supplemental oxygen, and he was the first climber to ascend all 14 peaks over 8,000 m above sea level. Furthermore Messner was also the first person to cross Antarctica and Greenland without snowmobiles or dog sleds. He also crossed the Gobi Desert alone. He wrote more than 80 books about his adventurous experiences. And he is also a nature conservationist and a political activist. Reinhold Messner is a sportsman larger than life, who finally came to understand the ecosophical principle of letting as many species flourish as possible.

[10] “Der Grenzgang beginnt dort wo die Show afhört” (Messner).

> 'Everything is subjective,' you say; but even this is interpretation. The 'subject' is not something given, it is something added and invented and projected behind what there is.—Finally, is it necessary to posit an interpreter behind the interpretation? Even this is invention, hypothesis.
>
> In so far as the word 'knowledge' has any meaning, the world is knowable; but it is interpretable otherwise, it has no meaning behind it, but countless meanings.—'Perspectivism.'
>
> It is our needs that interpret the world; our drives and their For and Against. Every drive is a kind of lust to rule; each one has its perspective that it would like to compel all the other drives to accept as a norm (p. 267, §481).

In the following, I will first reiterate the main theoretical incentives of this approach (a philosophical style with a touch of agony) with a short retrospective of ecosophy and ascetology. Still adhering to the imperative that practical philosophy must be elucidated by examples, and honouring Nieztsche's idea that everyone has a preferential compelling perspective on so-called facts, I will concentrate on a limited sample of specific cases of lived-through agony in endurance sport practices. What makes a deeply suffering sportsperson tick?

7.2 Retrospective: Agonizing Ecosophy and Ascetology

Sigmund Loland in his casuistry often refers to sports that thrive on a variety of talents, such as stamina, perseverance, agility, explosiveness, motor ability, and tactical insight. He does so for good reasons, since putting a wide range of talents to the test in alpine skiing, ball sports and the ecosophically reformed sprint-running scheme Loland suggests, have a good chance of realizing the 'extended, ecological Self'.

There are more manifestations of sport under the sun, however, I argued. Consider endurance sports, such as long distance running, road cycling, open water swimming and, the classic Ironman-distance triathlon. A common denominator is their monomaniac character: repeating the very same movement over and over again, at a glance without the faintest notion of playfulness. The apparently 'boring' nature of these sports of long breath does not have to exclude ecosophical joy, however. Even more, once swimming, running, and cycling are automated—when they are, so to say "running over our spine" (Van den Berg 1973, preliminary note, my translation)—they can provide unique opportunities to experience the beneficial sensation of a holistic unification with nature.

Endurance sports even may lead to a Spinozistic *unio mystica* that precisely overcomes the automated nature of the horizontal daily drag. Training sessions, especially the lonely ones, obviously are more prone to this overwhelming feeling of joy and unification with nature, since they provide more room for mental ecosophical roaming. However, also contests on a razor's edge must not be excluded from potential ecosophical enlightenment. Loland acknowledges the benefits of strenuous training and hard work.

> Indeed, training can be part of the ecosophical approach to sport even if it is considered an instrumental activity. For example, the aim can be to reach skill levels that offer a deeper understanding of the possibilities that offer a deeper understanding of the relational body. Work is acceptable as a means to increased joy and perfection. Here, sport is similar to most other human practices, such as the practice of doing eco-philosophy: periods of monotonous, hard work are sometimes necessary to reach deeper insight and values (1996, p. 80).

In endurance sport, however, training hard is not only an acceptable but a necessary condition for ecosophical optimization. Loland refers to Naess's norm that "[s]kill development is an optimal balance between width and depth!", between "specialization" and "all-around skill development" (p. 80). I contend that in endurance sport width simply equals depth. Racing is enhanced training. Acquired skills certainly need maintenance, but the quintessence of long distance is zealous perseverance. Since they are performed over a variety of distances an surfaces and in the case of triathlon also different skills, endurance sports meet several criteria Loland suggests for reforming record sports. They, however, still stick to the loathed record paradigm of ever going faster, higher and stronger at maybe at quite some costs. I have argued that this does not have to be ecosophically problematic per se, however. Endurance athletes distinguish between 'fast'[11] and 'heavy' and therefore relatively 'slow' races.[12] This diversification to some extent already dilutes Loland's (quasi-) record setting paradox in favour of a perspective that lets more flowers bloom, and thus opens up for a wide variety of record-setting. What makes endurance sports even more ecosophical is the fact that trying to win a race is not relevant for the majority of serious endurance athletes.[13] Ending as high as possible in your age-group, improving your personal best, finishing a race *tout court*, overcoming personal agony rather than agonizing opponents and finding your own pace seem to be more important. It is the heroic self-oriented agonistic effort that counts, not just the petty spectacle of scoring goals or breaking records.

To provide some counterbalance to Loland's emphasis on elite sport and record setting, I referred to Peter Sloterdijk's plea for a renewed systematic asceticism, or 'ascetology'. I argued that such an all-embracing but at the same time monomaniac upwardly-oriented training theory is laudable from a societal point of view. While 'traditional' sport philosophy is usually still preoccupied with elite sport, 'sport' (or perhaps rather, 'physical exercise') also has become an almost mandatory health enhancing technique for the crowd. In the Netherlands, for instance, the government advises people to exercise at least 150 min a week, preferrably spread out over several days.

Sloterdijk argues that we should learn to revalue training as such, not just as a means towards ends: fitness, health, appearance, looks, prestige, let alone to satisfy modern buzz words like 'interaction' and 'communication'(p. 4). Extrapolating the

---

[11] E.g. the Berlin marathon with its wide lanes, lack of sharp turns and its flat course, especially designed for record-setting.

[12] E.g. the mentioned Norseman triathlon, which ends with an extremely heavy mountain 'run'.

[13] Non-elite Ironman distance tri-athletes may spend up to 30 training hours a week.

summons from the statue towards an endurance sport perspective, I argued that in order to learn to (re-)value the Sisyphean nature of the human condition, we have to be patient, attentive, persistent, and endorse what William James has called 'the strenuous mood'. To achieve this sense of fulfilment it takes a lot of '*askēsis*', ancient Greek for exercise. For Sloterdijk asceticism represents a wider range of historical manifestations of striving for excellence trough dedicated training practices. Our ascetic planet is inhabited by individuals who are constantly and relentlessly training themselves: from monks and Stoics to Kafka's hunger artist, armless violon players, professional cyclists and weekend warriors.

This emphasis on all sorts of training theories may seem self-focused, but it may also have a broader scope: we train ourselves to become better citizens, contributing to a just and sustainable society. We must change our lives and strive for a new horizon of universal co-operative training practices, in short: a 'general ascetology'. Since Pierre de Coubertin's introduction of the modern Olympics in 1896, the athletic ideal, which was predominant in the agonistic Greek culture, has taken the lead again: "a transformation best described as a re-somatisation or a de-spiritualization of asceticisms" (p. 27). This renaissance of exuberant physicality also gives room to enact our '*Vertikalspannung*', our vertical tension, our inescapable tendency to always try to run faster, jump higher, and become stronger: *citius, altius, fortius*!

This empathic revaluation of the body doesn't have to mean, however, that the rational soul is simply replaced by a life which simply relies on reflexes, as I have argued in the metabletic or historical phenomenological Chap. 4. There I argued that *physis* and *psyche* are two sides of a coin in the modern athlete. Of course it is rather obvious to associate a strong athletic body with a strong will, but, on the other hand, a strong will to work up oneself also allows for the asthenic and the disabled to upgrade their abilities, if only they are willing to put in substantial effort.

To reach the full potential of the transformative power for the better of humans, the (quasi-)Sysiphean dimension of the human condition has to be revalued a such, Peter Sloterdijk argues. The quintessential point of the almost inexhaustible human potential for pushing back frontiers is that we first of all have to become fully aware of our ascetic disposition. Once internalised, asceticism implies freedom and thus a full potential for change. For Sloterdijk ascetic practices are, though mediated, not just means to ends, such as prestige, health or even social interaction or communication. Well-performed and well-understood practices also must touch upon something essential that we have lost on our way towards technotopia: humble perseverance.

We have to realise that life is a highly repetitive effort to grasp unconcealed being as such, rather than a one-dimensional, tailor-made training programme for uncompromising record-setting or linear personal growth. The common denominator of the agonistic games so many people play is that they, in the end, hint towards the personalised heights of humanity. Once we have sufficiently invested in automated repetition, we can climb any mountain at our own pace. If performed in the true ascetic mood, *agon* has infinite ecosophical potency. I now will switch to a more in-depth analysis of the 'self-agonizing' nature of specific endurance sport practices and their deeper meaning. Simple means, such as bicycles, may lead to rich ends such as self-acceptance and ecosophical self-actualization.

Now for some imaginative high-ascetic casuistry.

7.3 Agonize Thyself

Endurance sports offer an ecological palette of explicit and implicit forms of competition, varying from winning a race or striving for an age group victory up to improving your personal best or even simply finishing a 'race' and competing with oneself rather than with opponents. In conjunction with the possibility for dramatic interaction with nature they may offer, this variety in specific agonistic weight of endurance sports has considerable potential to transform humans for the better.

In particular, mixed endurance sports such as duathlon ('run-bike-run') and triathlon (swimming, cycling, and running) can serve as a grindstone for sharpening the wider, 'ecosophical' understanding of the subtly varying manifold of manifestations of agony in sport. Ascetologically speaking, they consist of a mixture of repetitive strenuous physical activities that symbolise the evolutionary coming of age of humankind: from swimming and amphibian crawling on land to running and using high technology on two wheels.[14]Another argument in favour of opting for 'stamina sports' as a running gag, is that serious endurance athletes often show high levels of insight in the quintessential role auto-agony plays in their specific ascetic niche. Racing in long distance endurance is often racing against yourself. In this sense agony is an end in itself, rather than a hurdle to be overcome on the road to victory over others.

When it comes to competitive events as such, endurance sports are highly formalized. Competitors have to follow a certain route by precisely described and regulated means and comply to specified rules. In the case of a long distance triathlon this respectively comes down to: swimming with goggles, wearing a wetsuit if the water temperature is 78 degrees Fahrenheit/17 Celcius or lower, using a

[14]To make it more exciting and heroic, the evolutionary logical order, which would make running prior to cycling, has been swapped: a triathlon ends with a running marathon, during which extreme and often almost literal agony becomes visible in the facial expression of the suffering athletes.

safety helmet and a bicycle that meets the regulations,[15] cycling without drafting,[16] fixing flats yourself, not accepting food, fluids or help from bystanders, and, finally, running on any pair of running shoes. On the other hand endurance athletes always to some extent have to improvise, and thus potentially broaden their ecosophical perspective. This need for improvisational adaptations is obvious during long solo training sessions. But also during races endurance athletes have to deal with the unpredictable whims of nature, and the specific features of the racing course, in a more process oriented and co-existent manner. This means: coping with waves and unintentional kicks in the stomach from opponents during the swim leg, cycling on slippery roads with potholes, overcoming sudden crosswinds, unexpected thunder-storms and false flats, finishing a marathon without getting dehydrated, and, finally, perhaps even glimpsing at the sublime in a "commodified world"(Gorichanaz 2016, p. 365).

To draft or not to draft?

On the one hand, long distance triathlons are regulated by the external institutional norms of the *International Triathlon Union*. On the other they are driven by

[15] Interestingly major technological breakthroughs in professional road cycling usually originate from triathlon. Probably the most well-known case is Greg Lemond's win of the 1989 Tour de France by beating Laurent Fignon in the final time trail with the help of triathlon aero-bars, which became common practice from then onwards.

[16] For all events with a cycling distance greater than 40 km, the bicycle draft zone is a rectangle 12 m long and 3 m wide that surrounds every bicycle competitor and every motorcycle on the bicycle course. The front edge of the front bicycle and motorcycle wheel defines the centre of the leading 3 m edge of the rectangle. Competitors have 30 s to pass through this zone. (British Triathlon Union).

the internal goods of the agonistic practice. They are characterized by a continuing subtle quest for finding the golden mean between contesting opponents and testing oneself, between appropriate aggression and the need for moderation in order to avoid muscular and mental damage.

Endurance sports neither simply reduce sporting activities to rule-bound, deontologically formalized[17] agony *per se*, nor to a sheer accidental means to conceptualize the world without an explicit notion of being faster, higher or stronger than other 'competitors'. The urge to agonize yourself until just before the tipping point, is a *conditio sine qua non* rather than an accidental epiphenomenon of the specific endurance practice. I will now submit a number of endurance experiences, described from the point of view of the practitioner, to an agonal analysis.

International Triathlon Union Powerman Long Distance Duathlon World Championships, Zofingen, Switzerland, September 4, 2012.

The race starts with a challenging 10 km run in the undulating woods around the charming little Swiss town. The first kilometer is steep uphill, so that the competitors are in the red almost immediately. The bike course is a hilly but achievable lap of 50 km, which has to be completed three times. The total height difference of the bike part is 1,500 m.

> But then you come to the 'pièce de la résistance', the second 30km run. I knew it was hilly. But I thought I'd got my legs reasonably used to running in the hills. However as I quickly realised doing 3 minute hill reps up the Gog magogs in Cambridge just doesn't cut the mustard when it comes to Zofingen. In my head I'd thought of it as 2.5km uphill, 10 km of more or less flat, then back down and repeat. Yeah right. In fact it's 2.5km of very steep climbing, followed by 10km of what I can only describe as a maze of sharp up and down, twisting pain at the top, followed by the 2.5km steep downhill and then repeat.
>
> I have no idea what we did at the top other than going up and down every possible hilly permutation on paths around the top of the hill. At no point could you get into a rhythm. The downhills were just as bad if not worse than the uphills. And it seemed to go on forever. The worst thing was knowing you had to do the whole thing twice. Whoever designed this course could not possibly have made it harder had they tried! It was like one of those nightmares I occasionally have if I'm stressed before a race when you're stuck in transition with no way out; a never ending web of sadistic pain.
>
> So no, I wasn't prepared for this run. Discovering the hellish 'maze' at the top was a bit of a setback to my race mentality and it took a whole lot of grit and determination to keep running! In previous Ironman races my legs have been tired at the end of the run, but not hurting. This time I wasn't so much tired as in agony. Every single muscle in my legs and gluts, even ones I never knew I had, was crying out to stop. As if that wasn't enough I also got a horrendous stitch for about 20 minutes (Gossage).

[17] Thou shalt not … a, b, c, etc.

Lucy Gossage at the hellish maze. (Source: Powerman)

Thus sighed the British du- and triathlete Lucy Gossage after ending as second during the powerman race. Of course she was trying to beat Eva Nyström, the Swedish powerhouse who in the end would beat her by 4 min and 22 s. But what Gossage also (or probably even particularly) was doing during the race, was overcoming muscle acidification, getting the very best out of herself and finding the balance between pain, perseverance and an occasional touch of pleasure. Gossage, who at that time combined being a professional triathlete with doing a PhD on kidney cancer at Cambridge University UK, was rather agonizing herself than fighting her competitors. She was suppressing pain and trying to find her own pace rather than keeping up with her opponents. This sense of 'auto-agony' seems to be crucial for long distance racing.[18]

> What you need at Kona is emotional control during and being able to race your own race.[19]
> Stick to your own race. Get what you expect out of yourself.[20]

[18] To my knowledge, in long distance triathlon only Australian Chris McCormack is a well-known intimidator during races.

[19] Rookie Terrenzo Bozzone (New Zealand) in an interview the week before the Ironman World Championship on Hawaii (http://triathlete-europe.competitor.com/2014/10/08/video-terenzo-bozzone-on-lessons-learned (retrieved on October 9 2014). A race he didn't finish, by the way.

[20] Triathloncoach Mat Steinmetz in an interview the week before the Ironman World Championship on Hawaii http://triathlete-europe.competitor.com/2014/10/07/video-mat-steinmetz-talks-kona-kit retrieved on October 9 2014).

An often vented 'shallow' reductionist response to why endurance athletes voluntarily immerse themselves into agony, suffering and anguish is the suggestion of a 'flow'-experience,[21] or the so called 'runner's high', a colloquial term for a sudden euphoric feeling or boost of energy experienced during prolonged strenuous exercise. In popular scientific publications it is often suggested that β-endorphins are responsible for this state of physically induced euphoria. A solid scientific study, using PET scans combined with recently available chemicals that reveal β-endorphins in the brain, was able to compare the brains of runners before and after a run (Boecker et al. 2008). This study confirms a specific role of the opioid system in the generation of the runner's high sensation. No conclusion however can be drawn "with respect to which of these opioid receptors is dominating the opioidergic effects related to endurance training" (p. 2529).

Another neurobiological article, somewhat reluctantly, also corroborates the neurobiological reward hypothesis for endurance. Cursorial, endurance oriented species, such as wolves, cheetahs and humans, are evolutionary adapted specifically to run, whilst non-cursorial mammals, say rats and ferrets, are not.

> This study provides the *first* [emphasis added, rw] evidence that inter-specific variation in neurotransmitter signaling may explain differences in locomotor behavior among mammals. Thus, a neurobiological reward for endurance exercise may explain why humans and other cursorial mammals habitually engage in aerobic exercise despite the higher associated energy costs and injury risks, and why non-cursorial mammals avoid such locomotor behaviours (Raichlen et al. 2012, p. 1331).

Alltoghether there thus is hardly solid and cogent scientific proof for a purely physical runners-high. The German anesthesiologist Patrick Welsch therefore has suggested a philosophically more challenging explanation for the sought-after state of Zen-like happiness, seemingly making one forget the indisputable anguish of body and mind while accomplishing a long distance endurance challenge. In his study on psychological coping and the alteration of β-endorphines (the supposed marker for a runners-high) in 11 medium and high level amateur tri-athletes (10 male, 1 female) after finishing an ironman (long triathlon: 3,8 km swimming, 180 km cycling, 42,195 km running: 7 male, 1 female) and an ultra-long triathlon (double distance: 3 male athletes) were analyzed. He concludes that there is no satisfactory biological explanantion for a hormonal runners high.

> The examination of possible relations between β-endorphines (β-E) and their relative alterations (delta β-E), which were measured in the plasma by RIA, on one side, and dispositional and situation depending psychological variables, on the other side, does not show any substantial correlations between them.... According to our results coping of ultra-endurance stress is not influenced by β-E and changes of mood (1993, pp. 2–3).

According to Welsch, 'self-realisation'[22] appears to be more important than 'health' in his sample of serious endurance athletes. Even more, the observed and interviewed

---

[21] The mental state in which one is completely absorbed in what one does, proposed by the Hungarian psychologist Mihály Csíkszentmihályi (1975).

[22] Without the capital S and the exclamation mark Naess suggests for the truly ecosophical variant, or probably even just 'ego-realisation'.

athletes judge their final ranking as less important than attaining their personal goals (p. 34). It seems that self-fulfilment, trying to improve one's personal best or just finishing an extreme long distance race, prevail over winning. He also has found that dedicated practitioners of extreme endurance sports often seem to have problems with the creation of a 'healthy' self-image. He suggests that the mythical aureole of the accomplishment of an (double) ironman, the heroification of the ultimate performance due to innumerable training hours and self-overcoming and the "triumph of willing over exhaustion and pain is also appropriate to (temporary) satisfy narcissistic megalomania" (p. 35, my translation).

To paraphrase the Delphic oracle, Welsch's insights suggest that knowing thyself equates agonizing thyself.

7.4 Pragmatic Integration

Welsch points at the "necessity of integrative theories" (1993, p. 3) enabling to develop a psycho-somatic-social medicine that addresses the "whole person" (p. 46). Such an approach might reveal more of the inner urge to agonize oneself in seriously practicing endurance sports. A less biomedical, more experiential and phenomenological version of this plea can be found in what Steen Nepper Larsen refers to as the magnanimous sensation of the "'inter-being' between the bike, the cycling experience, the ocean of interpretations" (p. 30) and the resulting potency for personal growth. His phenomenological approach overcomes the reductionist endocrinological hypothesis of the endorphin-induced 'runners high' and related notions such as notions of 'flow', 'mindfullness', Chi-running and other 'holistic theories'. By focussing on the cycling subject as such, Larsen tries to bridge the gap between 'continentalism' and a more pragmatic take on the wider implications of the vertically oriented strenuous mood:

> Being human is to be changeable biology, equipped with a flexible and learning brain, said to be taking form according to the things we do, experience, and reflect on. In spite of different inevitabilities like the history of evolution and the law of gravity, human nature is not determined, nor is the human body simply a load. When the neuronal couplings that concern the bike are grounded and formed in the brain, the synapses are influenced by spatial interpretations, the sensual quality of rough rides, the linguistic exchanges in a group of tense riders heading towards the final sprint, the appraisal of new asphalt with no hole. This happens via the muscular conquests through which you become aware of the power you posess, the future performances, and the options to which the road of the world might invite you. We are doomed to search for meaning. We can't help ascribing sense to things and events while we swiftly move sitting on slim gel-filled saddles, calm and stoic on steel, aluminium, or carbon frames (2010, p. 28).

Also Scott R. Kretchmar's philosophical reflections on 'distancing' in sport—a kind of levitation from the tangible world, "which can be modified but which cannot be entirely avoided"' (1982, p. 6)—provide an impetus for a deeper understanding of the triumph of the will over exhaustion, without throwing the agonistic baby out with the bathwater. Kretchmar, one of the founding fathers of the contemporary academic philosophy of sport, suggests that the seemingly everlasting pursuit for pushing the limits of pleasure and pain by athletes of long breath might be a necessary tool for distancing oneself from the 'tangible' everyday world, where 'convenience' seems to be the *Leitmotiv*. "Perhaps, in some fundamental sense, to be 'at a distance' from the surrounding milieu is to be humanly powerful and free" (p. 6). This idea of levitation seems especially applicable to endurance sports.

Also several literary authors point in the direction of distancing as one of the main drivers for uptaking the evolutionary habit of running or walking humans. Philosopher, novelist and runner Dirk van Weelden (2003) suggests that the calculating mind is on vacation during running. This enables it to overcome limits. Every run gradually evokes a sense of triumph. Every time the runner is able to demonstrate his 'enlarged' body in the open air is a victory over the domesticating forces beleaguering him. "These are the forces that attempt to make him slow and stiff, heavy and gloomy, afraid and quiet" (p. 57, my translation). Also Non-Nobel prize winner[23] and long distance runner Haruki Murakami runs to get away from the daily grind. During a run he simply runs, and tries to create emptiness around him. Of course it is impossible to totally block the psyche from entering any thoughts, he contends. Still, these mental intruders always remain subordinate to the emptiness of the run (2010, p. 27). While Alan Sillitoe's *The Loneliness of the Long Distance Runner Runner* (1959/2007) tells the story of Smith, a poor working class Nottingham teenager from a dismal home, who has bleak prospects in life and few interests beyond petty crime. In institutionalised Borstal the boy turns to long-distance running as a method of both an emotional and a physical escape from his situation.

> Because you see I never race at all: I just run, and somehow I know that if I forget I'm racing and only jog-trot along until I don't know I'm running I always win the race…
>
> Then he turned into a tongue of trees and bushes where I couln't see him anymore, and I couldn't see anybody, and I knew what the loneliness of the long-distance runner running across country felt like, realizing that as far as I was concerned this feeling was the only honesty and realness there was in the world and I knowing it would be no different ever, no matter what I felt at odd times, and no matter what anybody else tried to tell me. (pp. 42–43)

Philosopher and runner Mark Rowlands likewise points at the autotelic, non-material 'distancing' merits of running:

> Living amidst the darkened of the world, our lives are marred by the inability to recognize intrinsic value when we encounter it. Our lives are lived doing one thing for the sake of something else. Three score years and ten, or twenty, of an endless for-the-sake-of-which; decades of chasing what is valuable but only rarely catching it. To be in contact with something that is important for its own sake, and not merely for the sake of something else, would be to end this chase, at least for a while. For a time at least, one does not chase value, one is immersed in it (2014, p. xiii).

As for cycling, also cyclist-writers such as Peter Winnen (2000) and De la Cruz et al. (2016), who holds a degree in philosophy, confirm the occasional possibility of being at a distance from the daily grind as a benefit of their harsh métier.

[23] On 6 October 2017 the BBC "delved into the dejected world of his long-suffering fans whose sole desire—to see the Japanese writer win the prize—is thwarted every year."

However enticing the 'distancing' perspective may seem, there must be something more to explain the agonistic depth of endurance. As set forth in in the previous chapter, Hochstetler and Hopsicker (2012) have undertaken a precise systematic pragmatic attempt to indicate what I would call 'the specific auto-agonistic weight' of endurance sport. Or: The typical striving to agonize oneself in an extreme manner without the prospect of considerable prize money or fame. For them, the blessings of the Jamesian strenuous mood explain the disproportionate strains of the serious endurance athlete. Though the authors sympathise with the idea that probably anyone who participates in endurance sport applies for the possibility of growth through commitment, they reason that those who really immerse themselves in the specific endurance practice have a significantly higher potential for personal growth. There is a fundamental difference between occasional 'wide' joggers and weekend riders and dedicated 'deep' runners and cyclists. Truly dedicated athletes have heightened chances for a fully flourishing life and experiencing the depth of their auto-agonistic ascetic humanity; to season life and driving it into a meaningful corner.

Thus, beyond distancing from the daily drag, there is also the positive possibility of reconnecting to what is really important in life in an almost Heideggerian attempt to repair our forgetfulness of being. The reconnecting hypothesis suggests that fanatic cyclists and runners push their limits not to (just) satisfy Welsch's conjecture of a somewhat sadistic 'narcissistic megalomania' as the supposed driver for athletes to exhaust themselves. Serious endurance athletes also strive for creating a meaningful and prospering life by putting dedication, stamina and perseverance to the toughest of tests. The deeper the practicing, the higher the chances for Naess's celebrated Self-realization!

> Through an investment in the practice, runners and cyclists slowly take on the characteristics of the practice community and slowly develop an identity that is central to the way they see themselves and how others view them. Further, these endurance athletes build a coherent narrative of lived experiences, with training programs dedicated to improved performance, and a life story full of encounters while on the move (Hochstetler and Hopsicker 2012, p. 121).

As already argued, there may be a downside to Spartan dedication and the servile obedience to training schedules, carved in the sporting subject over decades, however. The monomaniac repetition, the eternal recurrence of the same automated movements seems to be the very opposite of the playful and voluntary character that sport is supposed to incorporate according to Johan Huizinga's *Homo Ludens: A Study of the Play-Element in Culture* (1955) and Bernard Suits's *Grasshopper: Games, Life and Utopia* (1978/2005). Endurance is not very lusory. Resilience and simply pushing on set the tone.

Nonetheless, referring to Johan Cruyff: to every disadvantage there is an advantage. Unlike ball games, sprinting and all sports that require highly developed motor skills and fast fibres, endurance sports do have special benefits to offer, also for the ageing athlete, Hochstetler and Hopsicker contend. There is a pitfall however. Whereas youth equates to agility and carelessness, coming of age from a physical perspective usually implies less flexible limbs, and more need of physical care, punctuality, rigidity and meticulous devotion, and therefore the danger of 'mechanized' practices. Also Rowlands (2014) makes an interesting, corroborating observation in this respect. Whereas the younger athlete usually runs in a Spinozistic way—that is: light, thoughtless, going up in the moment—the ageing athlete is prone to become a Cartesian mechanist, in the sense that the mind tricks the body by ignoring discomfort and pain.

But also the objection of reflexive mechanisation—meaning: being meaningless, since just running over the spine— should be nuanced. As argued in the metabletic Chap. 4, the paradox of endurance sport consists of the idea that automated reflexes in the long run may just enable moments of enhanced reflection. Also Hochstetler and Hopsicker acknowlegde this potential openness for meaning in the strenuous mood. Sticking to your training programma in this respect is not 'mechanic' per se, but potentially pays off in terms of human flourishing. It may be true that an overdose of structure limits individual autonomy, a precisely delineated programme can also be an asset in modern times. A well designed training schedule may be the very antidote to the penchant for overtraining of many endurance athletes,[24] and thus enhances the potential for ecosophical flourishing and meaningful encounters.

7.5 Into Deep Agony

As already mentioned in the previous chapter, in a subsequent article Hochstetler and Hopsicker (2016) point at the potential perils of monamaniac endurance and somewhat tone down their plea for the ruthless and solistic strenuous life, however. To flourish in the full Aristotelian sense, endurance athletes should also take care of their social environment, they now argue.

Although this social turn in itself is comprehensible and even defendable, one might as well argue that the very loneliness of the long distance athlete has the bigger potential for augmented ecosophical joy, and, paradoxically, also Self-realization! By means of a solistic vertical effort away from the wide but shallow horizontal dimension of the puny crowd, the endurance minion may tap into the deeper meaning of life. A voluntary stay in ice and high mountains, a look from a distance, ideally results in a heightened susceptibility for matters of ecosophical importance.

What still is lacking in Hochstetler's and Hopsicker's pragmatic analysis of the strenuous mood is a deeper introspective philosophical understanding of the quasi-Sisyphean nature of endurance sport. How can the lonely agonist, far ahead of the pack, be happy? To rephrase the agonistic paradox: why do serious endurance athletes commit themselves to painful activities? Is this the case because this voluntary stay in the pain cave helps them to reach for the sublime heights of human experience (Gorichanaz 2016)? Why do they fight against themselves rather than against

[24] I learned from an endurance trainer that difference between football-players and triathletes is that the first sub-species has to be incited when it comes to training, whilst the second always has to be slowed down.

opponents? Why do they sometimes slide into masochism? And what can we learn from this in order to further elaborate the ecosophical-ascetological perspective? What is there to gain in the monomaniac agony of the individual endurance athlete?

To further the issue, another agonistic observation has to be taken into account. Whereas Lucy Gossage's strivings discussed above can still be interpreted as a somewhat mild manifestation of auto-agonism, the following example, by all means, cannot.

The Dutch professional cyclist Gert Jan Theunisse won the mountains classification and the mythic Alpe d'Huez etappe in the 1989 *Tour de France*. He probably would have won more stages in *La Grande Boucle* if he would not have been such a lone wolf, who often rode solo when it probably would have been better to stay in the safe and warm womb of the peloton (Cf. Wielaert 1996). In terms of ranking he would have done far better if he had not been a rider who tested himself rather than his competitors. Many have even argued that he even could have won the *Tour*, the most prestigious bicycle race on earth, if only he had been a bit more rational.

Gert Jan Theunisse in the polka dot yersey (leader of the mountains classification in the Tour de France). (Source: Twitter)

In 1990 Theunisse tested positive in the *Flèche Wallonne* as well as in the *Bicicleta Vasca*. Although he argued that his blood dyscrasia was the result of a natural anonmaly, he entered the legion of suspected and contaminated cyclists, and due to a ban his track record came to a temporary standstill. In 1995 he abandoned the second stage of the *Tirreno-Adriatico*, and stopped his career as a professional road cyclist after receiving medical advice for heart trouble. At the end of the same year he became interested in mountain biking. The discipline of the mountain-bike riders and the feeling of being one with nature, while cycling off the beaten track, felt like a rebirth on two wheels to him. Despite his fragile health condition, Theunisse started to compete in regional mountain bike races. He managed to do quite well in his new discipline, in fact.

On 8 September 1997 misfortune struck. During a treacherous descent he was severely hit by a car. He was diagnosed as having an impairment in the motor and sensory function of the lower extremities. Initially he was unable to walk but step by step he recovered and returned to coaching and also training, and finally even racing. In January 1999, he won a mountain bike race but was not able to walk for 3 days afterwards. In June 1999 he was struck by a heart attack, from which he soon recovered, however. In 2000 he broke the *omerta* of the peloton, and admitted using doping during his career.

Theunisse then moved to Mallorca, where he started riding his mountain bike again, at least 150 km a day. He won the European over-30 championship in 2002, and kept competing at a serious level until 2005, in spite of consistent pain due to his paraplegia, having difficulty with walking straight and even sometimes suffering from involuntary muscle or spastic attacks. Although he was declared 13% functionally handicapped, during this period he nevertheless still was able to add 12 wins to his record of achievements.

In 2006 misfortune struck again. Theunisse was cut off by a Spanish truck. During the following argument the drivers, who turned out to be drunk, broke his knee. But Theunisse slowly recovered, and rose as a rider once more. After a long period of revalidation he started cycling again, all things considered the ultimate mainspring of his troublesome but challenging life. In December 2013, at the age of 50, after a period of heavy chest pain and fainting spells, he ended up in the intensive care unit and had a pacemaker implanted. Retrospective medical research showed that he must have suffered multiple heart attacks over the past years. Theunisse literally had been in agony: "Nobody understands why I'm still alive, they say that I must have had a whole peloton of angels on my shoulders." (Het Parool 2013). He now, against all odds, announced his retreat from cycling, literally his *raison d'être*. But in this specific case of agonistic masochism one of course never can tell.

But still, what's in it for the greater whole?

7.6 Enduring Agonistic Ecosophy

During an interview Gert Jan Theunisse once was asked why he kept going on with cycling, despite all his misfortune and physical discomfort. He answered, frowningly: "Pain is just a detail"(Creteur 2010, p. 124).[25] While for many people serious

[25] Cfr. for a detailed literary account of Theunisse's innate urge of overcoming necessary obstacles Jan Siebelink's *Pijn is Genot* (Pain is pleasure, 1992) and Jeroen Wielaert's *Gert Jan Theunisse: De fiets, de Fiets en Verder Niets* (The bicycle, the bicycle, and nothing else, 1996).

pain is a reason for quitting any physical challenge, Theunisse has been able to reduce this phenomenon to an inconvenience that can be ignored and suppressed in favour of some beckoning perspective. However, since this perspective for Theunisse was neither victory, nor prestige nor prize money: What was it that made him tick in such an extreme and apparently senseless manner?

There are befitting psychological explanations for Theunisse's deviant cycling behaviour. He was the kind of adoscelent that pressed cigarette-ends against his underarm to test his pain barrier. However, still following Welsch's call for an integrative theory that overcomes the reductionist soma-psyche watershed, there must be more at stake than temporary satisfaction of irrepressible impulses. But then again, what exactly are the benefits of the interminable repetition of the same old pedal and breast stroke or numb running pace? What makes the injuries and pain they often cause bearable?

Attempting to further integrate the respective strengths of phenomenology, hermeneutics, and pragmatism, I opt for what Van den Bossche has titled a "Heideggerian pragmatism" (1995, pp. 128–141).[26] This post-ontological and "meta-philosophical relativist" synthesis (2001, p. 44) enables us to challenge the prevailing word view of humans as sporting for either just prize money, playfulness or health, namely by "philoponically"[27] elaborating the holistic idea of (endurance) sport a preferred manifestation of "the art of living" (Van den Bossche 2010, p. 137).

My main line of reasoning can be summarized as follows. What is the good life? Or: how are we to live? Since ancient times the answer to this question usually is that we must work on ourselves and improve ourselves by way of training. My practical and practiced philosophical investigation has focussed on one particular dimension of this striving for human perfection by means of asceticism: endurance sports, such as long distance running, cycling and triathlon. These are all sports that flourish because of dedicated training, while talent is less important. This makes endurance sports not only accessible, but also increasingly popular among the sedentary crowd.

Especially the phenomenon of cycling has brought endurance sport within reach of the masses. Almost everyone can ride a bicycle, a high-tech artefact which is extremely energy efficient. This makes the bicycle a straightforward economic tool for a more sustainable lifestyle. But the stakes of a life that is to be fully lived in endurance are higher. How can endurance sport at large and cycling in particular contribute not only to self-knowledge, but also to self-improvement and to sustainability? Because of its competitiveness and decisively agonistic characteristic—at first sight the very opposites of peaceful sustainable coexistence—sport usually has

[26] Van den Bossche (1995, p. 128) argues that pragmatism is not exclusively an Anglo-Saxon case. He contends that Heidegger's philosophical toolbox also accommodates pragmatic chisels and pliers. Cfr. also Okrent (1988), Drabinski (1993) and Pollock (1995) for other attempts to integrate the phenomenological-hermeneutical tradition with pragmatism.

[27] "[T]he love of *pónos*, namely effort, burden and strain" (Sloterdijk 2009/2013, p. 194).

a negative connotation in environmental philosophy or 'ecosophy', a term which has been coined by Arne Naess in the 1970s and applied to sport by Sigmund Loland in the 1990s.

Inspired by Peter Sloterdijk's analysis of humans as an upwardly oriented training animals, and insights from historical phenomenology (or 'metabletics'), hermeneutics and pragmatism, I then argued for a vertically challenged life: a flourishing but tough existence on the rack, as in William James's 'strenuous mood'. In other words: serious hard pushing and sweating, rather than just pleasantly pedalling around a bit. This results in an upwardly oriented ecosophical life, leading to qualitative growth, human flourishing, durability and a change for better. Agonistic sport and environmental sensitivity: the twain can not only meet but may even merge.

Sloterdijk argues that an assessment of the pervasive and often polluting technosciences of the present has to build on a broad historical and cultural understanding of the ways in which science and technology have played a crucial role in the coming-into-existence of human beings. His *You must Change your Life* entails a broad interpretation of asceticism, and paves the way for a synthesis of the ideal-typical vision of sport as a purely self-referential activity and more flexible interpretations of sport as a truth- or knowledge seeking activity.

Whereas the 'traditional' analytic philosophy of sport still often seems to focus on the playful, ludic character of its subject matter, Sloterdijk points out that it is the ascetic, or 'laborious' inner nature of sport that should be taken into (re-)consideration in order to understand life at large via a philosophical analysis of agonistic sport and ecosophy.

Crucial in endurance sports is the state of agony, the intense anguish of body and mind of the practitioner, I argue. This almost masochistic self-inflicted pain appears to be more of an intrinsic *conditio sine qua non* than a necessary but insufficient condition which unfortunately has to be taken into account on the often bumpy road to victory, the improvement of a 'personal best', or just finishing a race. The question is if and how this largely inwardly oriented and sublimated aggression in endurance sport can be reconciled with values that are to be preferred in times of severe ecological problems. Can we reach beyond the call for a noncommittal "benevolent anthropocentrism"—putting as little pressure on eco-systems as possible while sporting, but not questioning the deeper meaning of sport as such—suggested by Meinberg (1991, p. 132) when it comes to sporting humans and fragile ecosystems?

Especially the heroic and ultra-ascetic *métier* of professional cycling seems to be infected with immorality. Doping scandals continue to dominate. Sloterdijk, however, argues for overcoming our obsession with the hyper-commercial professional regions of sport and to catch sight of the dedicated broader crowd.

> One must admit, however, that those parts of the sporting world closest to the 'circus'[28] in the ancient sense, especially in the vicinity of the Olympic industry and in the professional

[28] Cfr. footnote 3 for a more nuanced view on the idea of 'bread and circuses'.

> segments of football and cycling, have meanwhile become subject to a result fetishism that absolutely rivals the compulsive product-oriented thinking of the economic sphere. But what does this mean if, on the other hand, statistics show that in those sports there are ten thousand amateurs or more for every professional? (2009/2013, pp. 212–213).

For Sloterdijk road cycling is the modern variant of ancient heroism. Especially the vertically challenged lonely climber, who leaves all mortals far behind, deserves our respect as the contemporary substitute of the ancient god-athlete. Nepper Larsen furthers Sloterdijk's message of a 'cycling Pan-theism'. To him, riding a bicycle implies a myriad of personalised agonistic ways to explore the heights of humanity, with the ascent *hors catégorie* in stormy conditions as the *summum bonum*.

Still, what about the ascetological-ecosophical benefits of deeop agony? In *Agon in Nietzsche* Yunus Tuncel argues for a renewed Greco-Nietzschean 'agonistic individualism'. This sovereign Nietzschean loneliness, this voluntary exposure to icy mountains may be a counterpoint for the contemporary need for excitement of the masses, the petty spectacle of scoring goals. Other than sportspeople who strive for externalities—fame, prestige, winning at all costs, prizes, money—those who learn to agonize themselves in order to optimize their ascetic potentialities may become better human beings in the end. As already argued in the previous chapter, agonism may not be the ultimate response to human suffering, however, as Tuncel argues, "it is one response", one that "promotes strong individuality, the best response to human suffering" (2013, pp. 256–257).

The endurance sport practices I have described do meet several criteria for the positive *Umwertung* (revaluation) of *agon* towards self-knowledge and self-acceptance Tuncel is looking for. As a personalised form of *askēsis*, performed in the great, wide and preferably hilly open spaces at your own pace, endurance sport can overcome the 'ecosophical' tension between *agon* and ecology. That is, by revaluing the quasi-Sisyphean concept of *askēsis* as a contemporary and relatively clean and green means to fight the struggle with and against oneself, on the collective level it may help to overcome the association of sport with environmental damage and transform the human condition into a 'coexistent agony' wíth the world. This renaissance of a forgotten or neglected inclusive life-style[29] should take place, however, without throwing the competitive baby out with the bathwater. One can ride a bike to try to win a race or to improve one's personal best, vent one's *thymos* in agonism, maintain one's embodied mind, or move around in a more sustainable manner.
The epilogue of this study, entitled *Turning in the Widening Gyre* will indicate how to imagine a happily cycling Sisyphus, who finally manages to roll his personal boulder up the hill.

---

[29] Of course one may question if such a peaceful and coexistent bucolic lifestyle that Rousseau argues for ever has existed, or that the history of mankind better can be characterized as a Hobbesian *bellum omnium contra omnes*. But, similar to Van den Berg's metabletic method, this doesn't seem to matter in Sloterdijk's vision. It is the beckoning *perspective* in the sense of Nietzsche that beats the measure. It is possibility that makes a man or a woman tick.

Sisyphus at work: Il faut travailler, toujours travailler!" (Source: Clipart)

References

Boecker, H., T. Sprenge, M.E. Spilker, G. Henriksen, M. Koppenhoefer, K.J. Wagner, M. Valet, A. Berthele, and T.R. Tolle. 2008. The runner's high: Opioidergic mechanisms in the human brain. *Cerebral Cortex* 18 (11): 2523–2531.

Clarysse, W., and S. Remijsen. Undated. *Ancient Olympics*. http://ancientolympics.arts.kuleuven.be/eng/TD019EN.html. Accessed 7 Mar 2018.

Creteur, J. 2010. Gert-Jan Theunisse: "Pijn is maar een detail". *Wielermagazine* 17: 124–129.

Csikszentmihalyi, M. 1975. *Beyond boredom and anxiety: Experiencing flow in work and play*. San Francisco: Jossey-Bass.

De la Cruz, J., J. Gomez Pena, P. Horrillo, et al. 2016. *El Afilador Vol. 1: Articulos y cronicas cyclistas Gran Fondo*. Bilboa: Libros de Ruta Ediciones S.L.

Drabinski, J. 1993. Radical empiricism and phenomenology: Philosophy and the pure stuff of experience. *The Journal of Speculative Philosophy, New Series* 7 (3): 226–242.

Fukuyama, F. 1992. *The end of history and the last man*. New York: Macmillan.

Gezondheidsraad. 2017. *Beweegrichtlijnen 2017*. https://www.gezondheidsraad.nl/nl/taak-werkwijze/werkterrein/preventie/beweegrichtlijnen-2017. Accessed 23 Feb 2018.

Gorichanaz, T. 2016. Beautiful and sublime: The aesthetics of running in a commodified world. *Journal of the Philosophy of Sport* 43 (3): 365–380.

Gorris, L., and D. Kurbjuweit 2008. Philosopher Peter Sloterdijk on the Tour de France: "The riders are just regular employees". Der Spiegel Online International, July 2008. http://www.spiegel.de/international/europe/philosopher-peter-sloterdijk-on-the-tour-de-france-the-riders-are-just-regular-employees-a-565111.html. Accessed 5 Feb 2018.

Gossage, L. www.lucygossage.com. Accessed 15 Sept 2014.

Het Parool. 2013. *Pacemaker voor Gert Jan Theunisse*, 12 May. http://www.parool.nl/parool/nl/11/SPORT/article/detail/3556911/2013/12/05/Pacemaker-voor-Gert-Jan-Theunisse. Accessed 15 Jan 2015.

Hochstetler, D., and P. Hopsicker. 2012. The heights of humanity: Endurance sport and the strenuous mood. *Journal of the Philosophy of Sport* 39 (1): 117–136.

———. 2016. Normative concerns for endurance athletes. *Journal of the Philosophy of Sport* 43 (3): 335–349.

Horsley, J., and M. Horsley. 2013. *There's always pizza!* Thornville: J. & M. Horsley.

Howe, L. 2008. Remote sport: Risk and self-knowledge in wilder spaces. *Journal of the Philosophy of Sport* 35: 1–16.

Huizinga, J. 1955. *Homo ludens: A study of the play-element in culture*. Boston: Beacon Press.

Kirsch, A. 2015. Against cynicism: A philosopher's brilliant reasons for living. *New Republic*. http://www.newrepublic.com/article/113387/peter-sloterdijks-philosophy-gives-reasons-living. Accessed 5 Feb 2018.

Krein, K. 2014. Nature sports. *Journal of the Philosophy of Sport* 41 (2): 193–209.

Kretchmar, S.R. 1982. 'Distancing': An essay on abstract thinking in sport performances. *Journal of the Philosophy of Sport*. 9: 6–18.

Loland, S. 1996. Outline of an ecosophy of sport. *Journal of the Philosophy of Sport* 23: 70–90.

Lopez Frias, J. 2014. Internalism. In *The Bloomsbury companion to the philosophy of sport*, ed. C. Torres, 362–364. London: Bloomsbury.

Lunt, D., and M. Dyreson. 2014. A history of philosophic ideas about sport. In *The Bloomsbury companion to the philosophy of sport*, ed. C. Torres, 17–37. New York: Bloomsbury.

McNamee, M.J. 2012. Spirit of sport and the medicalization of anti-doping: Empirical and normative ethics. *Asian Bioethics Review* 4 (4): 374–392.

———. 2014. The spirit of sport and its place in anti doping policy. In *Companion to drugs and sport*, ed. J. Hoberman, V. Moller, and I. Waddington. Abingdon: Routledge.

Meinberg, E. 1991. *Die Moral im Sport, Bausteine einer neuen Sportethik*. Aachen: Meyer & Meyer.

Messner, R. *Reinhold Messner – Einblicke*. www.reinhold-messner.de. Accessed 14 Apr 2018.

Müller, A. 2004. Sports and Max Scheler's concept of death: High risk sports from an existentialist view. In *Movement: The art of life. Proceedings of international conference at Faculty of Physical Education & Sport*, ed. I. Martínková and M. Pelis, 53–65. Prague: Charles University.

———. 2008. *Risikosport: Suizid oder Lebenskunst?* Hamburg: Merus Verlag.

Murakami, H. 2010. *Waarover ik praat als ik over hardlopen praat*. Amstel Sport.

Nepper Larsen, S. 2010. Becoming a cyclist: Phenomenological reflections on cycling. In *Cycling philosophy for everyone: A philosophical tour de force*, ed. J. Ilundáin-Agurruza and M. Austin, 27–39. Chichester: Wiley Blackwell.

Nietzsche, F.W. 1901/1968. *The will to power*. Trans. W. Kaufmann and R.J. Hollingdale. New York: Vintage Books.

Okrent, M. 1988. *Heidegger's pragmatism. Understanding, being, and the critique of metaphysics*. Ithaka: Cornell University Press.

Pollock, A. 1995. *Pragmatism and hermeneutics: Rorty, James and Gadamer on truth*. Ottawa: University of Ottawa.

Raichlen, D.A., D.A. Foster, G.L. Gerdeman, A. Seillier, and A. Giuffrida. 2012. Wired to run: Exercise-induced endocannabinoid signaling in humans and cursorial mammals with implications for the 'runner's high. *The Journal of Experimental Biology* 215: 1331–1336.

Reid, H.L. 2009. Sport, philosophy and the quest for knowledge. *Journal of the Philosophy of Sport* 36: 40–49.

Rosenbaum, A. 1950. Indian far from bashful at chow. *Los Angeles Times*, October 18: C3.
Rowlands, M. 2014. *Running with the pack. Thoughts from the road on meaning and mortality*. London: Granta.
Sillitoe, A. 1959/2007. *The loneliness of the long distance runner*. London: Harper Perennial.
Sloterdijk, P. 2009/2013. *You must change your life: On anthropotechnics*. Cambridge: Polity Press.
Suits, B. 1978/2005. *The grasshopper: Games, life, and utopia*. Peterborough: Broadview Press.
———. 1988. Tricky triad: Games, play, and sport. *Journal of the Philosophy of Sport* 15 (1): 1–9.
Tuncel, Y. 2013. *Agon in Nietzsche*. Milwaukee: Marquette University Press.
Van den Berg, J.H. 1973. *De reflex: Metabletische tegelijk maatschappijkritische studie*. Nijkerk: Calllenbach.
Van den Bossche, M. 1995. *Kritiek van de technische rede. Een onderzoek naar de invloed van techniek op ons denken*. Utrecht: Jan van Arkel.
———. 2010. *Sport als levenskunst*. Rotterdam: Lemniscaat.
Van Weelden, D. 2003. *Looptijd*. Amsterdam: Augustus.
Venables, S. 2013. Tourism has devaluated Everst. *The Telegraph*, May 23. https://www.telegraph.co.uk/news/worldnews/asia/mounteverest/10075415/Everest-devalued-by-tourism-says-Stephen-Venables.html. Accessed 2 Feb 2018.
Welsch, P. 1993. *Bewältigung eines Ultratriathlons: Physiologische und psychologische Aspekte extremer Ausdauerbelastung/Coping of ultra-endurance triathlons: Physiological and psychological aspects of extreme endurance exercises*. Dissertation MD, Aachen.
Wheeler, M. 2017. Martin Heidegger. *The Stanford Encyclopedia of Philosophy (Fall 2017 edition)*. https://plato.stanford.edu/cgi-bin/encyclopedia/archinfo.cgi?entry=heidegger. Accessed 28 Feb 2018.
Wielaert, J. 1996. *Gert Jan Theunisse: De fiets, de fiets en verder niets*. Utrecht: Kosmos.
Winnen, P. 2000. *Van Santander naar Santander – Brieven uit het peloton*. Amsterdam: Thomas Rap.
Wyld, H.C., ed. 1958. *The universal dictionary of the English language*. London: Routledge & Kegan Paul Limited.

Chapter 8
Epilogue: Turning in the Widening Gyre

Every time I see an adult on a bicycle I no longer despair for the future of the human race.

(H.G. Wells). Cited in Ilundáin-Agurruza and Austin (2010, p. 1).

Cycling is like church—many attend but few understand.

(Jim Burlant). Cited in Ilundáin-Agurruza and Austin (2010, p. 7).

Abstract After pondering over the practical feasability of this 'idealistic' study in sustainable endurance and positioning it in the current philosophy of sport, I will reflect on a life virtuously lived in endurance, especially of the cycling kind. Starting off with challenging gravity and the first glorious unaided pedal strokes in the parental backyard. The first time with no hands. Followed up by widening circles: conquering villages, boroughs, provinces, countries, continents. And then in retrograde order back to the home base again, cultivating the garden, in search of destiny and origin by means of oriented repetition. On city bikes, racing bikes, time trail machines, mountain bikes and trekking bikes. Paved and unpaved. Alone or in a small group. Commuting, travelling, racing, climbing, gasping. Or just riding. Trying to find an own pace, and understanding that the good ascetic life comes with un-motorized reflexive-reflective moving around. To bring the motto of *The Rider* to memory: "Non-racers. The emptiness of those lives shocks me" (Krabbé T, The rider. Bloomsbury, London, 1978/2016, p. 1).

Keywords Verticality · Widening circles · The good life · Cycling

R. Welters, *Towards a Sustainable Philosophy of Endurance Sport*, Library of Ethics and Applied Philosophy 37,
https://doi.org/10.1007/978-3-030-05294-2_8

8.1 The Second Coming

Professional cycling races such as the *Tour de France* usually begin with a 'prologue', an introductory stage, often a short individual time trial under 10 km in the historic centre of a city. A prologue is more of an appetizer, it enables the audience to see the cyclists from nearby and to get a glimpse of what is to be expected during the next weeks. Who looks strong and confident? Which rider is still recovering from the preceding races? What kind of technical novelties are introduced this year? After 3 weeks of hard labour the *Tour de France* traditionally ends with a mass-sprint on the *Champs Elysées*. Often this is a breath-taking scene, mostly without any impact on the final ranking, however. The previous longer time-trails and especially the hard mountain stages already have determined the final winner. *Les jeux sont faits*.

'Epilogue' comes from the Ancient Greek verb ἐπιλέγειν (*epilégein*), which means 'saying something in addition'. An epilogue is a conclusion in hindsight, a peroration of a speech, the final part of a play, an afterword. Although the term is unfamiliar in cycling, I still would call this final Chapter of *Cycling for Life* an epilogue: a last outburst of energy, a belch, an exclamation mark that accentuates the different stages of the previous ordeal. Just as in elite cycling, the present epilogue also will not influence the end result, however. As of the prologue it must have been clear that the lonely climber, cycling far from the indiscriminate pack, would be the eventual champion of this stage race. Still, this final stage pretends to be more than just a petty spectacle of rolling wheels. It picks up the pieces and suggests an itinerary for the good life—strenuous but doable, at least if one is willing to put in sufficient effort.

As may be clear by now: we must change our life in the Anthropocene. The pivotal question is, however: are we able and willing to return to a life lived in appropriate modesty, a relatively simple *Sitz im Leben* which takes a considerable physical effort? What is the societal feasibility of the preceding plea for cycling and running, rather than driving or flying? Are we collectively able to exchange our highly commodified and motorized life for a bio-mechanical one? Are we willing to switch to homeotechnology and anthropotechnics that also pursue appropriate flourishing for non-human species? Hereby a few literary ponderings that abet my ascetological-ecosophical argument for changing our lives for the better.

As of the beginning of the twentieth century, the Irish poet William Butler Yeats (1865–1939) exchanged the transcendental beliefs of his youth for a more realistic style of lyric. Nevertheless, Yeats, who was awarded the Nobel Prize for literature in 1923, kept being fascinated with spiritual imagery and cyclical theories of life. *The Second Coming*, written in 1919, in the aftermath of the First World War, is a perfect specimen of his preoccupation with life as spirally circling away from a key point. Picked up by the whirlwind of societal turmoil, the falcon drifts away from its original destination: the falconer's glove.

> Turning and turning in the widening gyre
> The falcon cannot hear the falconer;

Things fall apart; the centre cannot hold;
Mere anarchy is loosed upon the world,
The blood-dimmed tide is loosed, and everywhere
The ceremony of innocence is drowned;
The best lack all conviction, while the worst
Are full of passionate intensity. …
Surely some revelation is at hand;
Surely the Second Coming is at hand (Yeats 1920).

In the aftermath of recent terrorist attacks, the Brexit referendum in the United Kingdom, and Donald Trump's election as the 45th president of United States of America, the stanza "Things fall apart; the centre cannot hold" regularly has been cited (Ballard 2016). Stripped of its apparent religious connotations,[1] *The Second Coming* still has high declarative potential beyond this isolated line, I argue. It's sense of *Unbehagen* also brings the idea of feeling detached in the Anthropocene to the fore, the ubiquitous sense of anger and impotency in a "new geological era in which the human has become an Earth-shaping force while the Earth itself as humanity's one and only life support system is becoming increasingly instable and less human-friendly" (Lemmens 2018). By alluding to the possibility of a second coming, the poem also expresses hope for betterment. It forms an open ended script, that has to be filled with a positive turn in our frenzied little lives, "rounded with a sleep"(Shakespeare 1983, p. 19)—so that the earth may become a somewhat friendly place for all creatures, great and small. In this sense the hope for an active, manmade second coming, an era in which the falcon spirals down again, is in line with Sigmund Loland's revised outline of a flourishing ecosophy of sport and Peter Sloterdijk's plea for asceticism as a real existing alternative for our unsustainable lifestyle.

In the critical ascetological-ecosophical perspective I developed, *The Second Coming* implies a call for proper and sober action. Other than transhumanists Nick Bostrom[2] (2002, Bostrom and Cirkovic 2008, Bostrom and Savulescu 2009, 2014)

---

[1]Yeats was a 'protestant agnostic', who as a public figure and politician often stood up against the Irish catholic clergy. As a poet his overall spiritual attitude best can be coined as idiosyncratic. "Yeats often used Christian themes in his poetry, but that was not the only religion that influenced his work. In the famous poem 'The Second Coming,' Yeats clearly references the Christian ideas of the second coming of Jesus Christ. The language of this poem could be interpreted as a reference to the Christian Armageddon. Yeats's personal beliefs extended well beyond Christianity, however. He held to a complicated view of history and religion's place within it. As an Irish native, he was fond of the folklore and mysticism of his Celtic heritage. Religious references in 'The Second Coming' have to do with the idea that the time of a Christian domination of history in Europe was coming to an end. Yeats was living in the aftermath of World War I and the Irish War of Independence. Both reflected Yeats' impressions of the world coming to an end" (Ridder 2014).

[2]There is arguably also a strain of problem-shooting pragmatism in Bostrom's thinking. On his website he states that: "(t)he search for crucial considerations relates to what I call 'macrostrategy': the quest to uncover links between long-term outcomes for humanity and present-day actions. How profound and inescapable is our cluelessness about such matters? And among the available bets, which one has the highest expected value?" (https://nickbostrom.com/#bio). In the end, however, Bostrom seeks salvation in high tech human enhancement rather than considering a profound mitigation of human behaviour as such. His line of reasoning is linear and progressive rather than winding and ascetic.

and Julian Savulescu, who advocates the 'principle of procreative beneficence'[3] and 'genetic moral enhancement' (1999, 2001, 2002, Savulescu and Kahane 2009; Savulescu et al. 2011), I propose to rely on 'mechanic' old school practices first. Why not simply train ourselves to put less pressure on ecosystems with more or less 'conventional' means—say bicycles or running shoes—rather than resorting to high-technological fixes? Why not apply Occam's ontological razor[4] and look for simple means that result in rich ends, rather than complex high-tech means with highly unpredictable and un-controllable outcomes? Why not train ourselves to become physically and mentally[5] resilient instead of selecting embryos for supposed perfection? And, following the pragmatic imperative of consequential truth: What is the window of opportunity for agonistic ecosophical asceticism anyway?

Since we have taken over planet earth the ceremony of innocence seems to have drowned, indeed. From a cynically 'realistic' perspective, therefore, one must question the feasibility of the voluntary retrograde movement to a life that is less motorized, less polluting, less consumptive, less convenient, less horizontal, and less allo-technology driven. From the doggish, down to earth kynical position Sloterdijk already proposed in his *Critique of Cynical Reason* (1983/1987), and transformed into a full-blown ascetology in *You must Change your Life* (2009/2013), the possibilities for a relatively simple, sustainable, but flourishing life are abundant, however. The uncomplicated life is lying at our feet, waiting to be revivified. We have to take another look at the earth rather than gazing at the stars.[6]

---

[3] Savulescu argues that parents should be able to "select the child, of the possible children they could have, who is expected to have the best life, or at least as good a life as the others, based on the relevant, available information" (2001, p. 413). This comes down to choosing for the absence of genetic disease, but also for non-medical traits such as sex and intelligence.

[4] *Numquam ponenda est pluralitas sine necessitate*. (Plurality must never be posited without necessity). The wrongly to Occam attributed phrase *entia non sunt multiplicanda praeter necessitatem* (entities must not be multiplied beyond necessity) actually comes from the Irish Franciscan scholastic philosopher Johannes Poncius (John Punch), who described the principle as a 'common axiom' (*axioma vulgare*) of the Scholastics (Cfr. Crombie 1959). In its current usage Occam's razor refers to the logic-rhetoric principle of not introducing unnecessary 'things' to explain a specific phenomenon. Bostrom and Savulescu, however, argue that humanity cannot be saved without introducing new (homeo-)technology. By doing so, they introduce the 'myth of progress', the idea that society inevitably tends towards the introduction of new and ever more sophisticated technologies. Reverting to Heidegger's idea of *Gelassenheit* (1966), I still contend that the idea of an appropriate withdrawal with conventional means would be the most elegant solution to overcome the stalemate between human expansionism and the limited resilience of planet earth.

[5] Which are two sides of the same coin, I argue.

[6] Also Elon Musk's idea of the future colonization of Mars suffers from the myth of progress and the idea that the forthcoming anthropocenic catastrophe only can be solved by a technological fix. "You want to wake up in the morning and think the future is going to be great—and that's what being a spacefaring civilization is all about. It's about believing in the future and thinking that the future will be better than the past. And I can't think of anything more exciting than going out there and being among the stars" (SpaceX). Why not concentrate on planet earth first and considering the idea of appropriate withdrawal? Why not first try drive and fly less and cycling more, emitting less carbon, burning more fat, and using only ecosophically sound technology instead of colonizing space? It may be exciting to be among the stars, but life is only possible on a planet.

The question then is whether philosophy is the proper tool for the necessary ethical-ecological wake up call. Ten years before Sloterdijk's *You must Change your Life* Hub Zwart (1999) already stipulated that practical philosophy should be receptive to what Sigmund Freud has called *Ichumarbeitung* (p. 169), self practice, a willingness to work on yourself. One should keep in mind, however, Zwart argues, that self-work is *langwierige Arbeit*, a process of long breath, with ups and downs and many restarts. Philosophical 'methods' probably never will definitely solve the problems they engage with. Still, practical philosophers should be willing to subject themselves to 'self-work' and 'self-criticism': "Self-analysis is inherent in practical philosophy as a philosophical practice" (p. 170, my translation).

Following this line of practice-oriented reasoning, it should be possible to create a shared, relatively simple, and practicable ecosophical ascetology, restoring balance to our lives, to society, and to our relationship with the 'natural'[7] world, I contend. We must develop a sense of possibility, as the main character (and alter ego) Ulrich phrases it in Robert Musil's post First World War novel *The Man Without Qualities* (1930/1995):

> So the sense of possibility could be defined outright as the ability to conceive of everything there might be just as well, and to attach no more importance to what is than to what is not (p. 10).

Yet, in the long run, Musil's man without qualities remains indecisive about whether to act or not. Ulrich stays in the depressive and passive asthenic mode, so to say. In *The Myth of Sisyphus* Albert Camus, on the other hand, reasons that the labours of Sisyphus, rather than as a sadistic punishment for challenging the Gods, can also be interpreted as the very antidote to the absurdity of life. In this novel of the 1957 Nobel Prize winner, the battle of rolling the stone up the hill over and over again *as such* already suffices to live an engaged life.

To recall Rodin's advice to the undecisive poet Rilke in Sloterdijk's *You must Change your Life*: go out, look for inspiration and start working again. *Travailler, toujours travailler!*

It is the idea that, one day, we may realise the (at first glance relentless) vertical challenge, against all odds, that keeps us going. Even more, I argue that after putting in a lot of training it may be quite possible to absolve a climb *hors catégorie*. At your own pace, this might be quite slow. But still, when you are really willing to push your limits you will succeed, even when you are just a mediocre dabbler. By quoting Pindar at the beginning of his collected essays (1942/1991), Camus also points at the feasibility of the strenuous ascetic life: "O my soul, do not aspire to immortal life, but exhaust the limits of the possible" (p. 7).

At the brink of ecological doom at the beginning of the Anthropocenic twenty-first century, the practical philosophical stakes are extremely high. Let us therefore

---

[7]Which, as argued in Chap. 2, is best defined in a 'Spinozistic' way as dynamic, ever 'naturing', and meanwhile highly usurped and manipulated by *Homo sapiens*. Protagoras's dictum that man is the measure of all things: of things which are, that they are, and of things which are not, that they are not, comes with a huge responsibility, I argue. Man, the former runner, runs nature, hopefully increasingly in a homeo-technical, if not an appropriately withdrawing manner.

follow Camus's positive Sisyphean message and William James's idea that (pragmatic) philosophy also can be a means to an end—such as a more sustainable world—by underlining which concrete acts are to be preferred, so that 'the best' may regain their confidence and become passionate again. Instead of waiting for a second coming, we better push ourselves, sweat a bit, and revivify the Sisyphus that still lingers inside all of us. Let us hop on a bike of our liking, and try its ecosophical-ascetological potential.

8.2 A Nearly Perfect Tool

Again, the bicycle is not only a metaphor for the good life, it also is a really existing, and nearly perfect bio-energetic tool for bringing about the ecosophical-ascetological betterment we need. In *Energy & Equity*, published in the *poionistic*, quality-oriented metabletic year 1974, Ivan Illich already underlined the importance of a life conducted on two un-motorized wheels. From an evolutionary point of view, humans are proficient walkers or runners. On their feet humans carry 1 g of their weight by expending just 0.75 calories in 10 min, which, compared to other mammals, is very efficient. The invention of the bicycle has speeded up the energetic efficiency of *Homo sapiens* exponentially, however.

> Man on a bicycle can go three or four times faster than the pedestrian,[8] but uses five times less energy in the process. He carries one gram of his weight over a kilometre of flat road at an expense of only 0.15 calories. The bicycle is the perfect transducer to match man's metabolic energy to the impedance of locomotion. Equipped with this tool, man outstrips the efficiency of not only all machines but all other animals as well (p. 60).

Illich furthermore emphasizes that compared to motorized vehicles bicycles are also quite cheap[9] and extremely 'green'.

> They can get the benefit of technological breakthroughs without putting undue claims on the schedules, energy, or space of others. They become masters of their own movements without blocking those of their fellows. Their new tool creates only those demands which it also can satisfy. Every increase in motorized speed creates new demands on space and time. The use of the bicycle is self-limiting. It allows people to create a new relationship between their life-space and life-time, between their territory and the pulse of their being, without destroying their inherited balance (p. 63).

Four decades after Illich's moral wake-up call the bicycle is still alive and kicking. Whereas the sale of new bicycles has decreased over the last years—from 140,000 in 2007 to 938,000 in 2016 (Thole 2017)— in the Netherlands (17 million inhabitants) on average each person still owns 1.3 bicycles, which is the highest rate in the

[8] Up to 5 or 6 times on a racing-bike, I would say.

[9] As already mentioned in footnote 3 the Cervélo P5X costs approx. 15,000 euro's. However, one can buy a reasonable city bike for 400 euro's and a racing bike for 600 euro's. Also bikes need to be maintained, of course, but they do not need fuel and there is no road tax for bicycles.

world.[10] Dutch men and women cycle on average 880 km per year spread out over 250 to 300 rides (Fietsersbond 2018).

'Functionality' is the most often-mentioned reason for riding a bicycle. In a densely populated, flat and (because of its many cycling lanes) 'cycle-friendly' country such as The Netherlands, the bicycle often is the most efficient tool to get from a to b, and back, especially in the city. Furthermore, 10% of the population can be qualified as a 'sports-cyclist' or 'racer', whereas also sunny Sunday-afternoon cycle-trips and so-called 'cycling holidays' on trekking-bikes with panniers are still quite popular in the homeland of the bicycle.[11]

As already stipulated in Chap. 2, especially off-road cycling increasingly leads to tension in small-scale forests or other wilderness areas—dunes, heath, moors, water meadows—that are visited by so many nature-lovers. But also ruthless pelotons of road cyclists dressed in bright spandex and anarchistic city cyclists (ignoring traffic lights, cycling on the pavement, cutting off other road users) cause more and more friction, notably with car-drivers. However, due to the good old Dutch tradition of 'poldering',[12] in combination with sophisticated socio-geographical and physical planning, new arrangements are made to facilitate all road-users, including cyclists choosing the narrow or the broad path, paved or unpaved, in a group or alone.

Altogether the tide for cycling is still quite favourable, also on a larger geographical scale. There is hardly a major Western city that doesn't have a 'bicycle plan' and a 'bicycle policy' these days. Furthermore is has to be noted that bike-racing is not limited to traditional racing countries such as The Netherlands, Belgium, France, Italy and Spain anymore. The current elite peloton harbours racers from all over the world. And so does the growing peloton of amateurs and weekend-warriors—from Japan to California and from Norway to South-Africa.

To put it briefly, cycling is in the air, as a clean and relatively fast means of transport in an urbanized world, but also as an increasingly popular 'sport' that keeps people healthy in an aging society (Accell Group 2017). Thus, as already argued, the bicycle is not only a perfect tool for going around but also for going up the mountains of life. Combined with the popularity of long distance endurance sport

---

[10] There is a significant rise of so-called 'e-bikes' (electric bike) over the last years. In 2016 there were sold 271,000 of these (Thole 2017).

[11] From personal experience I notice, however, that there is hardly a rise in young people when it comes to cycling for vacation. The audience of the cycling bus that brings me back home from Southern-France, Northern-Spain or middle Italy each year is rather middle-aged, if not *pensionado*. Young people increasingly opt for cheap flight-tickets to exotic destinations. In this respect they do not differ from the average environmental philosopher, however, who often also flies all over the world to discuss his or her sustainability message with peers. As we all know, preaching is far more popular than practicing.

[12] A 'Polder' is land reclaimed from the sea or the river. The 'polder model' model of consensus decision-making, based on the acclaimed Dutch version of consensus based economic and social policy making. It also can be described as 'a pragmatic recognition of pluriformity' and 'cooperation despite differences'. In this sense The Netherlands is a Jamesian country: 'truth' has cash value, is revealed in its consequences, and is considered a matter of temporary and flexible consensus.

on the whole (Da Fonseca-Engelhardt 2013; Shoak et al. 2013; The Active Network 2013), the potential for maximal Self-realization! by means of agonistic ecosophical asceticism is immense.

Still, there seems to be a discord between (endurance) sport as a health enhancer for the crowd and elite sport: “[I]n the case of sport for all, the values were mostly public health and social integration, which were promoted in conflict with other values such as the one-sided focus of organised sport on competition, athletic records and Olympic excellence” (Eichberg 2009, p. 443). In this study I argued that one possibility to bridge the gap between the dragging mass and the gifted athlete is to further zoom in on ‘boundary subjects’[13] such as the non-elite runner, rider and du- or tri-athlete—the increasing peloton of anonymous finishers and mediocre age-groupers. What about their inwardly oriented masochism, celebrated without the slightest chance of success? Why do such people defy strong headwinds, face storm and icy rain, and climb mountains *hors catégorie*?

In Chap. 7, I reflected on the contention of the anesthesiologist Patrick Welsch that there is no straightforward proof for a hormonal reward for the suffering of extreme endurance athletes. This made him suggest that the “triumph of willing over exhaustion and pain is also appropriate to (temporary) satisfy narcissistic megalomania”(p. 35). I then brought Yunus Tuncel’s idea of agonism as an adequate response to human suffering to the fore. Referring to Nietzsche, he emphasizes the healing effects of agony for channeling cruelty and destruction, as well as learning to deal with loss and death. Tuncel furthermore reasons that “agonism promotes strong individuality, the best response to human suffering” (2013, p. 257). In this sense fierce agonistic competition becomes a purgatory for the dark side of life, so to say.

Altogether, these suggestions of ongoing ego-strengthening, self-empowerment, self-cleansing and aggression ventilation cover the issue of voluntary and unrenumerated long distance suffering fairly well. Nevertheless, throughout this study I have also argued for a still more positive interpretation of serious endurance and the strenuous life by proposing a broad view which also takes human flourishing more explicitly into account. So that during a painful endurance ordeal an occasional smile may occur on the lined contestant’s face.

As for the specific case of cycling, this holistic approach ideally results in what Peter Sloterdijk during an interview referred to something which reverberates with Spinoza’s notion of *hilaritas*, namely cheerfulness, merriment, or good humour. “On the bike, people become optimistic. Then, one experiences that willing and being able to are one in a spinning cycling pantheism. Then you want to scream out: it works, it works, it works” (Giesen 2011, p. 8).

This sense of joy after a first un-aided bike-ride is probably familiar to all of us. Keeping up *hilaritas,* the sense of unencumbered detachment throughout life is

[13] Keulartz argues that boundary objects (‘grensobjecten’) enable coequal coexistence of conflicting beliefs without the necessity of consensus or compromise. They can have different meanings for different communities, but they are still sufficiently robust to have a meaning that overcomes locality (2005, p. 22). The same goes for endurance sportspeople, I argue.

problematic however, philosopher Mark Rowlands argues in *Running with the Pack: Thoughts from the Road on Meaning and Mortality* (2014). He refers to Spinoza's idea that being free is acting in accordance with necessity. "When you're 'in the zone' you act without acting. What you do is a perfect match with what the situation requires" (p. 21). This usually occurs during youth. Think of a perfect shot or a sprint during which you don't even notice your feet hitting the ground.

As of a certain age, when joints become stiffer and life less unconcerned, this sense of total physical freedom becomes a different case, however. Especially when it comes to stamina, Rowlands concludes. "The freedom embodied in running distance is very different—it is not the freedom of Spinoza, not the freedom of youth" (p. 21). This is when the physical body and the non-physical mind start to separate.

> The key to building distance in the long run is the ability of the mind to lie to the body and be convincing. …This Cartesian phase, where one lies to one's body and so seemingly—but presumably erroneously—demonstrates one's distinctness from it, is just the first phase, the first *face*, of freedom. There is another entirely more interesting face to unveil: an old friend of mine who whom I shall meet again today, assuming I last long enough. But without wishing to endorse Descartes' more general views of the relation between mind and body, it still seems true that whereas the freedom of youth effaces the difference between mind and body, the freedom of running distance accentuates it (p. 23).

This paradoxical idea of still feeling free—or probably rather: feeling truly human—while enduring hardship and pain, helps us to come to terms with life. As argued in the previous chapter, this is why former pro Gert Jan Theunisse kept on racing despite his injuries, and countered his interviewer by saying that "pain is just a detail" (Creteur 2010, p. 124). Besides cultivating stamina, endurance sport is also about accepting transience, as expressed in Heidegger's idea of *Sein zum Tode* (being towards death), put forward in *Being and Time* (1927/2008).[14] Only the sense of finitude grants full meaning to life.[15]

Rowlands in this respect refers to the Ancient Roman philosopher Cicero (106–43 BC), who already argued that philosophers understand the art of dying "to the extent that he or she is someone who knows how to spend time with the mind" (p. 23). And so does the mature endurance athlete:

> A distance runner knows how to spend time with the mind—whether it does or, more likely, does not survive death. To run distance is not to run from old age; it is to run towards it. Far from a crisis, it is an acceptance of the point one has reached in life. And so the freedom of running distance is, it seems, the freedom of age. Far from reclaiming the freedom of youth, the freedom of distance running involves claiming, perhaps for the first time, an entirely different sort of freedom (pp. 23–24).

---

[14] "Dying is something that every Dasein itself must take upon itself at the time. By its very essence, death is in every case mine, in so far as it 'is' at all. And indeed death signifies a peculiar possibility-of-Being in which the very Being of one's own Dasein is an issue" (p. 284).

[15] This is the point where trans-humanism goes astray, I argue. A life without an end is meaningless.

It is this substitution of light-hearted juvenile merriness for ripe human flourishing that keeps the endurance athlete going.

And, again, especially the bicycle appears to be well suited for the long haul. By sparing joints over time inevitably becoming stiffer and more brittle, this extremely efficient bio-energetic tool magnifies the possibilities for life-long agonistic ascetic ecosophical joy. This makes a bicycle a nearly perfect tool for withstanding the rigors of time. Let us have a final look on how the diamond frame fits into the current philosophy of sport.

8.3 Philosophy of Sport at a Cross Roads

In Chap. 5, entitled *Ascetic Practices, Hermeneutical Cycles and Ecosophical Endurance*, I concluded that the philosophy of sport[16] still seems to be locked in a binary view. On the one hand narrow internalists, or formalists, argue that sports are uniquely constituted by their rules. This point of view can be referred to as the *autotelic* stance. In the 'analytic' perspective, sport is considered an end in itself, constitued by rules and (eventually) shared conventions on how to play or race well. On the other hand broad internalists, or interpretivists, reason that sport is more than just a gratuitous and playful end in itself. In this wider line of reasoning sport also can (or even should) be a means toward other ends. These may be less desirable ends, such as national pride, prize money, a ruthless quest for records, but perhaps also more positive ones, such as challenging the existing order, advancing international peace and contributing to a more sustainable world.

The latter, more hermeneutically, phenomenologically and holistically oriented *heterotelic* view has gained considerable ground in recent years. In 2016 an issue of *Sport, Ethics and Philosophy* was even solely dedicated to the case of *Hermeneutics and Sport* (eds. Lopez Frias & Edgar). There I argued for sport as an 'agonal' competitive social practice sport and as means to an inescapable end: seeking knowledge, understanding the human condition, cultivating virtue or bringing about the shift towards the more ecosophical-ascetological lifestyle we desperately need in the Anthropocene. Sport should not be reduced to navel-gazing rule-bound sheer 'ludic' physical activity. 'Sport' should become more of an all-inclusive way of life. A way of life which cherishes the upwardly oriented spiral effort of the marathon rather than the 100 m dash in the blink of an eye.

Also in *The Journal for the Philosophy of Sport* the tide is turning towards more asceticism and less ludism. The Canadian sport philosopher and former editor-in-chief John Russel in 2015 argued for considering resilience as *the* central virtue in sport (thus not 'playfulness' or a 'lusory attitude'). "Despite its importance it is an overlooked virtue in philosophy of sport and classical and contemporary virtue theory" (p. 159). Whilst also the Danish sport philosopher Kenneth Aggerholm has

[16] Academically starting off with the first issue of the *Journal for the Philosophy of Sport* in the poionistic metabletic year 1974.

undertaken a well-designated effort to link Sloterdijk's ascetological findings with a plea for a less result-oriented and more process-oriented virtuous approach in the philosophy of sport (2016). He distances himself from the internalist account of sport (p. 356), which generally still concentrates on specific games and races in favour of a more process oriented take. "I will even venture the suggestion that an ascetological account of sport would revise the primacy of competition to argue for the primacy of practising" (p. 357). To reinforce his provocative argument, Aggerholm applies Michel Foucault's 'Stoic' idea that engagement is "recognised, thought, lived, and practiced as a constant test" (Foucault 2005, p. 437) to the case of sport.

> Similarly, the role of practicing in athletic ascetics cannot be reduced to preparations as a contrast to competing in performances. An ascetological understanding of sport transcends this practical distinction and confirms what is well-known by practitioners in sport, namely that you can indeed be practicing whilst performing in competition[17] (Aggerholm 2016, p. 358).

To strengthen his acknowledgment of asceticism as the central virtue of sport, Aggerholm refers to Sloterdijk's plea for choosing the difficult vertical challenge instead of the easy way out, for instance by opting for the quick fix of using drugs.

> It is in this context that one must assess the future of modern sport. Like a Herculean collective, it is standing at a crossroads. Either the athlete continues to act as a witness to the human ability to take forward steps at the threshold of the impossible—with unforeseeable transference effects on all who involve themselves in the appealing spectacle—or they continue along the path of self-destruction that is already marked out, where moronic fans shower co-moronic stars with recognition from the very bottom, the former drunk and the latter doped (Sloterdijk 2009/2013, pp. 417–418).

Following Sloterdijk's vertical 'philoponical' imperative, Aggerholm argues for the primacy of practising over competition, thus paying attention to the active formation of habits. "This, I believe, can contribute to a more sustainable account of sport where competition and achievement is not separated from, but intimately related to, and indeed presupposes, practising" (2016, p. 363).

Russel and Aggerholm still reside in the realm of organized elite sport, however. They revalue resilience and asceticism above all for competitive purposes. They do so for good reasons, since the philosophy of sport often still reflects on canonised sport: high impact sport covered by the media. There is more under the sun, however. Cycling life is not just about 'inefficient'[18] bicyle racing, it is also about

---

[17] On the other hand, I argue, it is also possible to compete during training sessions, such as the among cyclists well-known phenomenom of sprinting for place-name signs, but also comparing efforts on specific stretches on the STRAVA website.

[18] In the narrow internalist perspective a bicycle race is inefficient, since it usually ends at the beginning, and one can furthermore think of more efficient (motorised) means to get around. To recall Bernard Suits's often cited definition of game-playing: "To play a game is to attempt to achieve a specific state of affairs [prelusory goal], using only means permitted by rules [lusory means], where the rules prohibit use of more efficient in favour of less efficient means [constitutive rules], and where the rules are accepted just because they make possible such activity [lusory attitude]" (1978/2005, pp. 54–55).

'efficient' bicyle commuting as a really existing alternative for car-driving and getting stuck in frustrating daily traffic jams. Fully-fledged practical philosophical cycling is not just about achieving a state of non-committal jolly playfulness, it is also an efficient, integrative and transformative means for the flourishing life. In this line of thick reasoning Marc Van den Bossche rejects the idea of sport as 'a self-absorbed universe' (2010, p. 133). A daily trip of 84 km to and fro the University of Brussels makes him survive examination periods by cleansing the mind and the air. "This is what I mean with sport as the art of living and as an integrated part of our world. The only world" (2010, p. 136, my translation). Also the almost obligatory ascent of the Mont Ventoux for every serious hobby cyclist should rather be a magical subjective experience than a race against someone else's clock, Van den Bossche claims. He cycles for life, so to say.

In *Cycling Philosophy for Everyone* (2010) sport philosophers and cyclists Jesús Ilundáin-Agurruza and Mike McNamee[19] also look at sport through a wide-angle lens. They argue that bike-riding (also) can be "a pedagogical path toward self-reliance and achievement where toy became a vehicle for freedom and confidence … to explore new territory" (p. 253). One starts wobbling, but once the turning of the pedals has been automated bicycling becomes as large as or even larger than our little life.

> In stage races prologues are short, intense efforts, preambles that set the tone for what's to come. Childhood is but one such dash. Before we know, it's over. Yet, our future passions are often moved by the muscular memories of the first pedal strokes that hold the secret reasons to our zest for both bike and life. Are we overplaying things? How can a bicycle, a triangle on two circles, as Pythagoras might conceive it, help us to live flourishing lives? (p. 253).

The authors contend that during late adolescence "idealism meets that lust for all kinds of action" (p. 254). During these formative years one learns "what doing something for love (of the game) is about. It marks the origin of a cultivation and appreciation for the practice of cycling" (p. 254). On their "breakneck descent down the tricky switchbacks of non-instrumental values and intrinsic interest" (p. 254) the authors argue for an embodied variant of the eudaimonic life.

> Attempting to articulate the value of that cycling can offer us on the way to living good lives—at all levels—is not easy but one necessary training element is getting clear on the nature of value and valuing.
>
> When we treat others—pets, people, possessions, even activities—as ends worthy in and of themselves, just for what they are, we value them intrinsically. They are not mere instruments. We take a spin 'just because'. … If all you really value your bike for is an external end, like getting to the shops quicker than walking or running, well, why not take the car? Wouldn't it be more efficient since it is faster and takes much less effort? But those who love life on two wheels don't ride this one-way road of efficiency-only thinking. And we all know that if all that matters is the results or the win, well shortcuts seem to be easily justified (p. 255).

[19] Both are also former presidents of the *International Association for the Philosophy of Sport*, which makes their line of reasoning an argument of authority, I would say.

One of these pejorative extrinsic 'shortcuts' that according to the authors easily comes to mind is doping (p. 255), a topic which often has been discussed in the *Journal of the Philosophy of Sport* and *Sport Ethics and Philosophy.*[20] Ilundáin-Agurruza and McNamee also point at meliorative heterotelic ends of the un-motorized triangle on two smoothly spinning circles: "cycling can be a wonderful means towards fitness, political and ecological goals" (p. 255). Roughly as of Sigmund Loland's above extensively discussed *Ecosophy of Sport* (1996), a fair amount of attention is given to ecological issues in both journals mentioned above.[21] This in particular relates to what I am inclined to call 'knowledge-oriented deep ecological issues', such as *Sport, Nature and Worldmaking* (Krein 2008). These 'ecosophical' ponderings ideally result in the process-oriented broad internalism Ilundáin-Agurruza argues for elsewhere.

> This leads to a consideration of 'thick holism', which advances a functional, flexible, and ecological model that emphasizes the continuities, fluidities and connections that permeate our body-mind. It holds that we are richly complex, dynamic, animate organisms whose bodily, psychic, emotional, affective, sensory, kinetic, energetic, socio-historical, and environmental facts form a unity that is consummated to various degrees of competence and integration (2014, p. 230).

Sloterdijk contends that modern sport is standing at a cross-road. Will it be an admirable striving for physical-moral excellence or a perfidy resulting in fetishism? Society at large approaches a similar junction. Will we move on and destroy the planet, or are we able to restrain ourselves and put in a physical effort to satisfy our horizontal needs?

Too much environmental philosophy is still focused on the thinking subject in the low-power mode. As if the good life is a matter of deep thinking rather than hard labour of a somewhat happy Sisyphus during his or her "voluntary pursuit of pain" (Elcombe and Tracey 2010, p. 243). The reflective person, fully aware of every single step, will probably only take a few hesitant ones, however: taking the bus to Machu Picchu, climbing a few stairs, and enjoying the breathtaking view. Yet, learning to automate your steps and pedal-strokes, and putting in a substantial and horizontal ascetic effort can take you much further. Say, crossing a continent on your own feet or un-motorized wheels, and running or riding all the way up to the highlights you meet en route, paved or unpaved.

The peloton has great aerodynamic advantages when it comes to speed, and thus range.[22] Even more, in *The Articulation of 'The We' in Bicycle Riding—Phenomenological Perspectives on Social Synchronization* (2012) Steen Nepper Larsen provides evidence for the claim that the runaway peloton is but a soulless

[20] Inserting 'doping' in the search engine of the websites of the journals resulted respectively in 83 and 129 hits on 18 April 2018.

[21] 23 Hits in *The Journal for the Philosophy of Sport* and 38 in *Sport, Ethics and Philosophy* on April 18 2018. Knowing that *Sport, Ethics and Philosophy* only exists as of 2007, this imposes the hypothesis that this journal is more susceptible for the thick take on sport.

[22] Which makes riding in a group more ecosophical, since it combines the hilarious joy of sensing speed with a wider scope.

machine, composed of cycling automats whose embodied actions run over the spine.

> Racing cyclists in a formation are alert creatures. They form a dynamic behavioral interaction in an ever changing flock. The cyclists do not just 'read' the landscape, the quality of the road, the density, coordinates and speed of the cars and other vehicles, and the wind, they also notice, interpret and anticipate each other's movements. They form a social organism on wheels in space and time; a 'social meaning that the cyclists ride, work, bond, and experience together. (p. 123)

When it comes to finding life, it is indeed advisable to go by thyself, by two or, go on then, three, as Jan Hendrik Van den Berg (1973) already argued in the metabletic Chap. 4.

> Only in a small group, or in solitude, one can find life. He, who is in search of life, leaves the broad road, which leads to doom and destruction, and chooses the narrow path, that demands full attention at every single step (p. 170).

Ilundáin-Agurruza and McNamee reason that the answer to the question how cycling can result in a flourishing life "must take on a narrative structure; an examination of the cycles of life and of life on cycles" (2010, p. 253). Let's have a thick narrative look at a few selective cycling experiences.

8.4 Lonesome Cannibal

At formidable speed he descends the Eyserbosweg in his rainbow sweater on his orange steel Molteni. He doesn't seem to notice me. He is in utmost concentration. At our house, about 200 m before the right-angle bend to the right, he still gets 80 km per hour, I guess. It is not disregard for death that makes him ride that fast. It is pure control, and he is heading for yet another victory.

The gap with the peloton is considerable, as usual. Big enough in any case for team leader Briek Schotte to kick the brakes. He has seen that I am washing my car with my dad, and he asks me to fill the water bottles of his superhuman *protégé*, the biggest cyclist of all time, not at all eclipsed by the lordship of the fallen angel Lance Armstrong. *Chef d'équipe* Schotte was the impatient deputy of God on earth. I still often fantasize about that (in my memory) quite sunny Saturday March 29, 1975,[23] when the peloton still descended our very own Eyserbosweg (road to the forest of the small Dutch village Eys) during the cycling classic *The Amstel Gold Race*. As of the early 1980s during the *Gold Race* the peloton went the other way round and started climbing the 900 meter long and on average 11.5% steep climb, with an intersection of 17.1% on the last 200 m.

[23] Memory is a poor counsellor. I found out later that this was one of those typically grey Dutch days.

Weekend warriors *en masse* ascending the Eyserbosweg

Large mosquito orchid, observed near the Eyserbos

What would have happened if I had not filled Eddy Merckx's drinking bottles with the lime-rich water from the neighbouring water extraction area the Roodborn, located in the beautiful, orchid-lacquered brook valley of the Eyserbeek that flowed from our tap? Would he have collapsed during the final? Would he not have won the Gold Race, and therefore only have had a record of 524 race victories instead of 525?

Eddy Merckx, the man who always attacked, and very often won, the all-devourer, with the nickname 'the cannibal'. The *Übermensch* with the shiny black hair. The sovereign with the long, muscled legs, always sitting super concentrated on his bike. With a look that seemed to be stuck somewhere between anger and determination.[24] The ultimate cyclist, the man with the ideal spirit in the perfect body, who often cycled ahead of the pack and who never gave it away. Winning was his *Leitmotiv*, on the flat and in the mountains. Collecting as many *maillots jaunes* and *maglias rosas* as possible.

At that time the germ must have been planted for becoming an observing character. Participating without reserve, merging into a group, just acting on the spinal cord, I have always had my reservations. Or, as the philosophical psychiatrist Jan-Hendrik van den Berg 2 years before Eddy's dizzying passage along my parental home in response to Rousseau's attempt to let the 'me' dissolve into the *moi commun* already stated: "The moi commun scratches itself, where the flea bites. The moi commun stretches the leg, if this leg gets a slap under the kneecap" (1973, p. 117, my translation).

As indicated earlier, the metabletic prophet of doom provides an escape clause for the highway to perdition: leave the broad road leading to destruction, and choose the narrow, unpaved path that requires attention at every step. Van den Berg doesn't seem to understand the true nature of cycling, I argued. Precisely by optimally spinalizing the pedal, the cycling subject acquires an excellent opportunity to escape the masses and to find his or her true, often lonely destiny. Cycling is a vital driver for true experience, rather than a hindrance. The Eyserbosweg was paved in 1975, but also quite narrow and not always easy to mount for mechanised traffic, witness the sign with the inscription 'Blocked during snow and ice'. On such a road you better jump away from the peloton, which Merckx did so diligently when he passed me at the age of nearly 13. I picked up his message to move ahead of the masses.

Beyond the close encounter with Merckx, memories of my youth consist of pictures of 'natural objects' north, east and west of my parental home on the Eyserbosweg. As of my tricycle years I gradually enlarged my little world via widening concentric circles. Although I soon managed to ride on two wheels, it took quite a while before I climbed the Eyserbosweg by bicycle all the way. I went to school on foot. That was mandatory because it was simply too dangerous for children from the surrounding hamlets to go into the valley with a bicycle with a coaster brake.[25] I was only allowed to de- and ascend the part of the Eyserbosweg until our home when I was sent out for groceries in the village or to train or play at the local football club. But my red single-speed with thick tires was quite heavy, and actually a bit too big. I was already happy when I managed to absolve the first 200 m to our house with an average grade of about 8% without getting off.

The first time I thought I would conquer the remaining 700 m of the climb was during a planned individual time trial with two friends and a cousin from the entrance of our garage located under our house. Trajectory: Eyserbosweg uphill, four kilometres down via the hamlets Eyserheide and Trintelen, descend the

---

[24] I apparently did not notice that the wear and tear already was beginning to show (Cfr. the metabletic Chap. 4).

[25] Dutch bikes often do not have hand-brakes, since these are not necessary in an on average flat country.

Trintelerberg, turn right at the church and then, to complete the circle, ascending the 200 m of the Eyserbosweg to our house. Unfortunately, my mother stopped me just when it was my turn. Too dangerous. Frustration. Pouting inside. I must have been 11 years old.

When I went to high school I got a new bike; one with three gears, which came in handy over the slightly undulating trail of 10 km that I had to go back and forth five times a week. On my new bike, on the way back from school, I sometimes passed our house and I always went up bit further. At some point I must have reached the summit without getting off. But, strangely enough, that joyful event has largely faded from my memory. Apparently the way up itself has been more impressive than reaching the actual summit itself: the entrance to the apple orchard of farmer Grooten on the right, the 18 Holstein cows of farmer Hamers at the left, the badger castle in the high bank on the right halfway, and the chestnut tree just before the wood and the steepest section.

During these formative years I still often chose the unpaved alternative way up through the meadows to the forest (actually, there was not even a trail), looking for orchids, lizards and sundew. But gradually the bike took possession of me, especially its vertical capacity. Because we only had a room antenna, I went to see the Tour de France at the neighbours. Flat stages, I did not have much with that. The lonely climber with the painful gaze, somewhere on the flanks of the Izoard or the Tourmalet, expansions of the Eyserbosweg as it were, I was seized by that.

About a year after I had filled Eddy Merckx's water bottles during his memorable descent from the Eyserbosweg, our quiet Eyserbosweg was suddenly plagued by big hordes of cyclists who wormed themselves up during the weekend. This inspired me, although I didn't like the scale. After a heavy ankle injury in football I was allowed to buy a road bike after the plaster had been removed: a lilac *Motobécane* of 499 guilders, with ten gears. The membership of a cycling club didn't cross my parent's mind. I was not sorry about that either. That nervous twist around the church in a peloton and the accompanying massive falls did not attract me. I just wanted to go out and conquer a concentrically increasing world. Initially towards the Slingerberg in Geulle, where the NK took place regularly, and after the World Cup victory of Jan Raas in 1979, also to the Cauberg in Valkenburg, and then also towards the Ardennes, where the climbs were longer, the landscape more ominous and the number of touring cyclists considerably less numerous.

At the return of the ever larger circles I made, every now and then I still cycled up to the summit of the Eyserbosweg. Just to see if everything was still in its natural place. Around my seventeenth I traded my walking sessions through the meadows and the Eyserbos for running, which is faster and therefore more efficient. Even during my student days, when I covered the distance between my new hometown Nijmegen and Eys, some 160 km, a couple of times a year on ever more sophisticated racing bikes, I kept on running in the warm womb of my biotope of yesteryear—to keep in touch with my native soil.

According to Heidegger, unlike people often think, meditative thinking does not estrange us from reality. On the contrary, it just keeps us focused on the *hic et nunc* of our being, our 'existence'. To switch on meditative thinking again, Heidegger insists that we have to

> dwell on what lies close and meditate on what is closest; upon that which concerns us, each one of us, here and now; here, on this patch of home ground; now, in the present hour of history.
>
> Does man still dwell calmly between heaven and earth? Does a meditative spirit still reign over the land? Is there still a life-giving homeland in whose ground man may stand rooted…? (1966, pp. 47–48).

There is a simple but elegant possibility for keeping your feet on the ground, I argue. The boys from the village were obsessed with the desire to get their driving license as soon as possible after eighteen. I chose to cycle through and for life. I opted for a gently ticking chain, the buzzing wind, the smell of freshly cut grass over the chilly cocoon of the car. Gradually I became a total cyclist, riding a city bike, a mountain bike, a cross bike, a racing bike, a time trail bike, and a trekking bike.

For the last 25 years, during summer vacation, I cycle with my wife for at least a month with around 25 kg of luggage on a trekking bike. Especially in southern Europe, the Alps,[26] the Pyrenees and less well-known mountain ranges deeper in the Iberian peninsula. Ernest Hemingway, who in his French years was an avid cyclist, already stipulated that if you drive a car, a mountain pass is also impressive, but the memory of a mountain is only really adequate if you have tamed it by bike (Brouwer 2017).

The bicycle is a perfect tool for sensory mastering vertical ánd horizontal challenges. In 2001 during a sabbatical, we cycled for 6 months through the USA, Australia and New Zealand. Especially the coast to coast trip through the States recalls warm and tactile memories. Jean Baudrillard states in *America* (1986/1989) that the vast landscape that is so characteristic of large parts of the United States should ideally are experienced by car. This postmodern philosophical sociologist may be right when he contends that our European image of the States is mainly formed by road movies with Dodges and Buicks on cruise control through almost infinite lunar landscapes.

> All you need to know about American society can be gleaned from anthropology of its driving behaviour. That behaviour tells you much more than you could ever learn from its political ideas. Drive ten thousand miles across America and you will know more about the country than all the institutes of sociology and political science put together. (p. 53)

As argued, for the same reason Pirsig reasons that the motor-cycle is the preferential for coming to terms with the States. Yet, there is another way of understanding the land of the brave. My wife and I cycled from Virginia to Oregon, starting at the Atlantic coast, crossing the humid Old South, the arid Mid-West, the stunning and crisp Rockies, and ending up at the moist and raw Pacific shore. We experienced the

---

[26] Until the middle of the eighteenth century, the Alps were fearsome, a place that you had to avoid. The Mont Blanc was popularly called the 'montagne maudite'. In Romanticism, an Alpine trip becomes a sort of obligatory pilgrimage for every self-respecting writer or intellectual (Lemaire, 2002, p. 87).

US countryside in a way that provides ample scope for un-filtered sensory input and meditative thinking. As of St. Louis we largely followed the trail of Lewis & Clark, who in 1804 made the crossing of the Rocky Mountains to the west coast, partly on foot, partly on horseback. 200 Years later the Lewis & Clark trail is nicely paved boulevard to the West. This makes the crossing of the Rocky Mountains a lot easier of course. Still, riding a bicycle at an appropriate pace leaves sufficient time for truly experiencing and sensing country, land and earth (Pesses 2007).

On the bicycle I managed to broaden and deepen my horizon of experience, my frame of reference, my view on flourishing life. Gradually even a certain feeling of 'poionistic' superiority towards non-cyclists and even non fanatic cyclists has settled in me. I have to admit, however, that during races I often am busy with countable things—the *poson* side of the spectre. These are exactly those things that according to Van den Berg and Heidegger make current times so cold and spiritless. What initially evoked my admiration as an unselfish and wordless phenomenon—the Eyserbosweg and its ever increasing hinterland—meanwhile also has been reduced to coordinates, gradients, graphs and tables. Or as Umberto Eco's *The Name of the Rose* ends: *Stat rosa pristina nomina nomina nuda tenemus*: the rose of yesteryear exists only as a name, all that remains are naked names (1980/1984, p. 502).

At my better moments, however, I am still occupied with what I used to do at home at the Eyserbosweg: drifting away from and returning to a fixed point, a 'residence', to speak with Heidegger. Even after 35 years and now that my parents have not been living there for a long time, I every once in a while cycle around or run in this landscape that is still the measure of all things for me.

8.5 Cycling for Life

As said, as a reasonable runner and ditto cyclist, for my competitive aspirations I finally opted for the duathlon, or the run-bike run. On September 4th 2009 I participated in the World Championships Powerman in Zofingen, Switzerland[27] for the first time. This race boils down to running 10 km, cycling 150 km and finally running another 30 mountainous kilometres. The bicycle lap has to be covered three times. It is not allowed to draft during cycling. The breaking point is the Bodenberg, whose summit is at 755 m. Just before that summit is the worst part: 200 m at 16%.

At the third ascent of the Bodenberg 'my pump was empty', to use one of these typical Dutch cycling expressions, meaning: totally exhausted. At a bunch of Swiss spruce trees I noticed the after-image of a black-haired courageous cannibal in a rainbow sweater on a steel Molteni. As if he wanted to say: simply push on, Sisyphus, you'll manage. In spite of my extremely narrowed consciousness, I saw the parallels: a similar gradient as the final part of the Eyserbosweg, comparable road

[27] I already described this race through the eyes of pro-racer Lucy Gossage in the previous Chapter. I probably even suffered more than she did.

conditions, the same green meadows, Holstein cows, a few scattered houses, a silent tree, a bunch of flowers. For a moment I thought I was home again.

Since this was a serious race I did not take the time to get off the bike and explore the surrounding meadows, looking for an orchid or a tree frog. Still, every time I descended the Bodenberg I had a glance at the powdered peaks of the Berner Oberland, a *mysterium tremendum et fascinans,* an unpaved, sublime, almost unreachable[28] vertical summit. For sure, the higher the ascetic stakes the narrower the ecosophical perspective. But in its totality, cycling is the ideal option for experiencing place dynamically.

A child cycles his or her first laps around the yard. Relieved from the auxiliary wheels, it slowly learns to defy gravity with a helping hand in the back. The first parental release is the ultimate moment of happiness for the cycling subject, who now discovers the blessings of the repetitive pedal-stroke. What follows is the episode of slowly leaving the familiar environment in concentric circles, alone or in a peloton. During the secondary school time, the first longer cycling tours are undertaken. Adolescence is the time of finding your way to the sun and freedom. In that phase of life not the circle but the line that moves away from the starting point is the favourite geometrical figure. Those who at the age of 18 do not exchange the bike for a 'grunting beast', as the Dutch novelist Ferdinand Bordewijk (1933/2012) calls a car, may go on with becoming true cyclists. Their development may differ from lonely home-work cyclists to weekend-warriors and highly competitive elite-amateurs. From dedicated marathon mountain bikers to world cyclists, traversing the continents for a good cause, constantly reflecting new horizons to the primal experience of the landscape from their youth.

When it comes to sustainability, at first sight walking seems to be the preferable option. However, in a world that has opened up, the walking range no longer suffices. The modern nature-hiker preferably flies to Corsica or Patagonia. The true world cyclist covers a wider green range, say by cycling from Nijmegen to Tierra del Fuego, at least as long as the Bering Strait remains frozen during winter time.[29] In this manner the bicycle can make a considerable contemporary contribution to fighting the inconvenient Anthropocenic truth that awaits us.

Beyond these ecological considerations, bicycle riding also is a perfect tool for dialogically (re-)connecting with nature and physicality. Van den Bossche argues that hikers and bikers both are prone to sensory input, be it at a different state of matter.

> In a walk through nature you feel a form of physical connection and there is a dialogue as it were. The hiker sees nature, the cyclist experiences nature. These are two different forms of dialogue. Thanks to the dialogue one comes to an understanding of things. And that dialogue happens through the game of question and answer. The better your fitness, the higher you set the threshold, the longer you move almost casually through the landscape.

[28] Qoud non. Daring downhill mountainbikers do, after being dropped off by a helicopter.

[29] Since this is not very likely in times of rapid global warming: there are also plans for a tunnel or a bridge from Russia to Alaska.

> My bicycle has also become a walk in this sense: the effort no longer overcomes the pure enjoyment.
>
> I want to see cycling as part of a hermeneutical framework. Sitting on the bike you listen to an environment. Your body is all ear (2006, pp. 27–8, my translation).

This will sound familiar to every experienced cyclist. From a certain level onwards, once you've reached the state of quasi-thoughtless *souplesse,* you are able to enter into some kind of multi-sensory dialogue with the environment. Of course this is easier when the road is flat and well-paved and the wind is low, but also the toughest of vertical challenges still leaves some elbow-room for ecosophical enrichment of the narrowed senses. The former Dutch pro Peter Winnen even thinks that the cyclist is capable of deeply experiencing the landscape, even better than the walker. He was asked if he actually had taken notice of the beautiful Alpine landscape during his winning ascent on Alpe d'Huez in 1981.

> I was not believed when I answered: "Everything that escapes the average tourist."
>
> Anyone who ploughs through the landscape on a bicycle does not only see it, but also gets it through the legs. There is more memory in my legs than in my head. Cycling is a matter of thinking with the legs (2000, p. 234, my translation).

> I look at my unwashed legs. They are lying here on the bed in front of me. I think of the treasures that are stored in it. Every meter I have ever driven is registered in it.
>
> In detail, as if never one second has passed, I find everything in it. Every country that I have visited, all the landscapes that I have ever crossed, all cities, all beauty and all the dirt. They were both my super-sense and my motor, endowed with a colossal and infallible memory. Each cell separately. Every muscle, tendon, is divinely stupid. What do you think about this: every cell a universe. Eternity is within reach (p. 237, my translation).

In the same manner—be it at a lower competitive level—I sucked up the Eyserbosweg area in my limbic system. The grounds of my youth have become an archetype, nested in every single body cell. In that sense, every bike ride that I complete is an 'oriented repetition', after a term by the Dutch philosopher, anthropologist, and organic farmer Ton Lemaire. He reasons that in a globalizing world, the space in which we move so hyperactively has become a homogeneous web of knots.[30] All points in space might be interesting, and thus worth the visit.[31] At the same time, or perhaps precisely because of this, there is an urgent craving for familiarity. "In this way man succeeds in maintaining himself in the infinity of time and space through celebration and concentration of his actions" (1997, p. 105, my translation). This longing for the oriented repetition of what is deeply known, is what makes a falcon spiralling down to his falconer again. Lemaire finely expresses that these sweet *lieux de mémoire* are constitutive for meaningful life.

> That is why the landscapes of our youth can accompany us for a lifetime and call up an indefinable homesickness when we see them again or recall them. The space in which we live for a longer period of time becomes an animated space, places in it acquire a special

[30] This similes Van den Berg's critique of spinalism. Other than Van den Berg, however, Lemaire sympathizes with Rousseau.

[31] Many people can no longer indicate a specific country or capital on a map, but they know how many flight hours it takes to get there.

> meaning because we attach ourselves to them and they have become part of ourselves in the long run; if they change, something inside us is touched at the same time. …
>
> A person who grew up in the north of the Netherlands will usually long for the vastness of that landscape with the grand skies and distant horizons, while someone whose youth played in the mountains would find it difficult to get used to in the flat and low country of the Netherlands (2002, pp. 41–42, my translation).

In the mountains that I now climb, mechanized but not mechanically, I still discern the afterimage of the primeval hill of my youth. When I cycle in the Italian Alps, I also look for that wild chestnut tree that marks the beginning of the steep stretch of the Eyserbosweg. In the Cévennes I search for the 18 Holstein cows of farmer Hamers. And in Northern Spain I am looking for the bright red poppies from the high graves opposite the house in which I grew up. On a regular basis I can still experience the rose of yesteryear as a rose. And not just as a naked name. There is still life out there. You can find it when you put in some effort. Hop on your bike, widen your gyre, spiral down again, and cycle for sustainable life.

References

Accell Group (The Premier Cycle and Fitness Group). 2017. *Annual Report 2016*. http://annualreport2016.accellgroup.com/docs/Accell_AR_2016/index.php?nr=22&language=en&r_code=Accell_AR_2016. Accessed 19 Mar 2018.

Active Network. 2013. *State of the industry: 7 trends in endurance particapation*. June 13. http://www.activeendurance.com/blog/2013/06/13/7-trends-in-endurance-participation/. Accessed 19 Mar 2018.

Aggerholm, K. 2016. On practising in sport: Towards an ascetological understanding of sport. *Journal of the Philosophy of Sport* 43 (3): 350–365.

Ballard, E. 2016. Terror, Brexit and U.S. election have made 2016 the Year of Yeats. *The Wall Street Journal*, 23 August.

Baudrillard, J. 1986/1989. *America*. London: Verso.

Bordewijk, F. 1933/2012. Blokken - Knorrende beesten - Bint. Amsterdam: Nijgh & Van Ditmar.

Bostrom, N. 2002. *Anthropic bias: Observation selection effects in science and philosophy*. New York: Routledge.

Bostrom, N., and M. Cirkovic. 2008. *Global catastrophic risks*. Oxford: Oxford University Press.

Bostrom, N., and J. Savulescu. 2009. *Human enhancement*. Oxford: Oxford University Press.

———. 2014. *Superintelligence: Paths, dangers, strategies*. Oxford: Oxford University Press.

Brouwer, E. 2017. Ernest Hemingway was een renner. In *De 141 beste Nederlandse en Vlaamse sportverhalen van 1945 tot nu nu*, ed. A. Van den Bogaard, 1121–1132. Amsterdam: Carrera.

Camus, A. 1942/1991. *The myth of Sisyphus and other essays*. New York: Vintage Books.

Creteur, J. 2010. Gert-Jan Theunisse: "Pijn is maar een detail". *Wielermagazine* 17: 124–129.

Crombie, A.C. 1959. *Medieval and early modern philosophy*. Vol. 2. Cambridge, MA: Harvard University Press.

Da Fonseca-Engelhardt, K. 2013. Participation and performance trends in ultra-endurance running races under extreme conditions – 'Spartathlon' versus 'Badwater'. *Extreme Physiology & Medicine* 2: 15. https://doi.org/10.1186/2046-7648-2-15.

Eco, U. 1980/1984. *The name of the rose*. London: Book Club Associates.

Eichberg, H. 2009. *Bodily democracy. Towards a philosophy of sport for all*. London: Routledge, Taylor and Francis.

Elcombe, T., and J. Tracey. 2010. Strechted elastics, the Tour de France, and a meaningful life. In *Cycling philosophy for everyone: A philosophical tour de force*, ed. J. Ilundáin-Agurruza and M. Austin, 241–252. Chichester: Wiley Blackwell.

Fietsersbond. *Fietsen in cijfers*. https://www.fietsersbond.nl/ons-werk/mobiliteit/fietsen-cijfers/. Accessed 15 Mar 2018.

Foucault, M. 2005. *The hermeneutics of the subject*. New York: Palgrave Macmillan.

Giesen, P. 2011. Peter Sloterdijk, Duitse filosoof: Naar de sportschool om de planeet te redden. *De Volkskrant* April 30, 2011: 7–8.

Heidegger, M. 1927/2008. *Being and time*. New York: Harper & Row.

———. 1966. *Discourse on thinking*. Trans. John M. Anderson and E. Hans Freund. New York: Harper and Row.

Illich, I. 1974. *Energy & equity*. New York: Harper & Collins.

Ilundáin-Agurruza, J. 2014. Nothing new under the sun: Holism and the pursuit of excellence. *Sport, Ethics and Philosophy* 8 (3): 230–257.

Ilundáin-Agurruza, J., and M.W. Austin, eds. 2010. *Cycling philosophy for everyone: A philosophical tour de force*. Chichester: Wiley Blackwell.

Krabbé, T. 1978/2016. *The rider*. London: Bloomsbury.

Krein, K. 2008. Sport, nature and worldmaking. *Sport, Ethics and Philosophy* 2 (3): 285–301.

Lemaire, T. 1997. *Wandelenderwijs*. Baarn: Ambo.

———. 2002. *Met open zinnen – natuur, landschap, aarde*. Baarn: Ambo.

Lemmens, P. 2018. *Thinking and imagining the Anthropocene*. Expert workshop with Timothy Morton. (Announcement). Nijmegen, 23 March.

Lopez Frias, F.J., and A. Edgar, eds. 2016. Special issue: Hermeneutics and sport. *Sport, Ethics and Philosophy* 10(4).

Messner, R. *Reinhold Messner – Einblicke*. www.reinhold-messner.de. Accessed 14 Apr 2018.

Musil, R. 1930/1995. *The man without qualities. A sort of introduction and pseudoreality prevails*. New York: Alfred A. Knopf Inc.

Musk, E. *SpaceX*. http://www.spacex.com/mars. Accessed 11 Apr 2018.

Nepper Larsen, S. 2012. The articulation of 'the we' in bicycle racing: Phenomenological perspectives on social synchronization. *Revista Portuguesa de Ciencias do Desporto* 12: 123–126.

Pesses, M.W. 2007. *Do two wheels make it more authentic than four? Spaces of bicyle tourism*. Master thesis, California State University, Northridge.

Ridder, K. 2014. Yeats' work: Influence of religion on Irish poet's writing. *Newsmax* 29 October. https://www.newsmax.com/fastfeatures/yeats-poems-christianity/2014/10/29/id/603990/. Accessed 14 Mar 2018.

Rowlands, M. 2014. *Running with the pack. Thoughts from the road on meaning and mortality*. London: Granta.

Russel, J.S. 2015. Resilience. *Journal of the Philosophy of Sport* 42 (2): 159–183.

Savulescu, J. 1999. Should we clone human beings? Cloning as a source of tissue for transplantation. *Journal of Medical Ethics* 25 (2): 87–95. https://doi.org/10.1136/jme.25.2.87.

———. 2001. Procreative beneficence: Why we should select the best children. *Bioethics* 15 (5–6): 413–426. https://doi.org/10.1111/1467-8519.00251.

———. 2002. The embryonic stem cell lottery and the cannibalization of human beings. *Bioethics* 16 (6): 508–529. https://doi.org/10.1111/1467-8519.00308.

Savulescu, J., and G. Kahane. 2009. Brain damage and the moral significance of consciousness. *Journal of Medicine and Philosophy*. https://doi.org/10.1093/jmp/jhn038.

Savulescu, J., R. Ter Meulen, and G. Kahane. 2011. *Enhancing human capacities*. London: Wiley-Blackwell.

Shakespeare, W. 1983. *The complete works of William Shakespeare*. London: Henry Pordes.

Shoak, A., et al. 2013. Participation and performance trends in ultracycling. *Open Access Journal of Sports Medicine* 4: 41–51. https://doi.org/10.2147/OAJSM.S40142.

Sloterdijk, P. 1983/1987. *Critique of cynical reason*. Minneapolis: University of Minnesota Press.

———. 2009/2013. *You must change your life: On anthropotechnics*. Cambridge: Polity Press.

Suits, B. 1978/2005. *The grasshopper: Games, life, and utopia*. Ontario: Broadview Press.
Thole, H. 2017. *De opmars van de e-bike in 7 grafieken*. https://www.businessinsider.nl/e-bike-elektrische-fiets-verkoop-rai-vereniging-bovag/. Accessed 4 Apr 2018.
Tuncel, Y. 2013. *Agon in Nietzsche*. Milwaukee: Marquette University Press.
Van den Berg, J. 1973. *De reflex: Metabletische tegelijk maatschappijkritische studie*. Nijkerk: Calllenbach.
Van den Bossche, M. 2010. *Sport als levenskunst*. Rotterdam: Lemniscaat.
Winnen, P. 2000. *Van Santander naar Santander – Brieven uit het peloton*. Amsterdam: Thomas Rap.
Yeats, W.B. 1920. The second coming. In *Michael Robartes and the dancer*. Dundrum: The Chuala Press http://www.potw.org/archive/potw351.html. Accessed 12 Mar 2018.
Zwart, H. 1999. Praktische filosofie als filosofische praktijk – en als teratologie. *Filosofie & Praktijk* 20 (4): 169–178 http://www.academia.edu/21789188/Praktische_filosofie_als_filosofische_praktijk-en_als_Teratologie. Accessed 27 Sept 2017.

About the Author

After obtaining a MA in philosophy at Radboud University Nijmegen in 1989, Ron Welters (1962) fulfilled his alternative civilian service as a researcher at the Christelijk Sanatorium Zeist. In 1992 he became a free lance journalist. As of 1996 he started working for Radboud University, first as an employee of the *Studium Generale*, and later also as a teacher. Until the summer of 2015 he taught courses in the Philosophy of Sport, Food Ethics, Ethos of Science, Scientific Writing, Science Journalism, Science & Media and Effective Communication of Biomedical Research. As of 2016, together with Prof. Dr. Marjet Derks, he teaches *History and Philosophy of Sport* at Radboud University.

In 2009 he became an active member of the international sport philosophy community. As of this time he has been working on this dissertation as a so-called 'external Phd-student'. Over the years he presented some ten papers and organized several scientific discussions on (endurance) sport and environmental issues. Since 2014 he is a non-delegate member of the board of the *European Association for the Philosophy of Sport*. In this capacity he organized the yearly conference in Nijmegen, his hometown, in 2017. For every attendee there was a bicycle available. What else.

Ron Welters is a dedicated age grouper in run-bike-runs and all kinds of running and cycling events. He is a three time finisher at the Powerman Zofingen, Switzerland ('the toughest duathlon in the world'), and a two time bronze medal winner in his age group during the national duathlon-championship in The Netherlands. He doesn't own a driving license and spends his holidays on a heavy loaded trekking bike with Nicolet, preferably in high mountains. Recently he also discovered the blessings of trail-running, during which every single step still demands his full attention.

R. Welters, *Towards a Sustainable Philosophy of Endurance Sport*,
Library of Ethics and Applied Philosophy 37,
https://doi.org/10.1007/978-3-030-05294-2